Prepare for any design challenge with CGP!

You just can't fluke your way through AQA GCSE Design and Technology — the new Grade 9-1 course will really test your mettle. And your metal.

Luckily, this CGP Revision Guide explains everything you'll need to know with no-nonsense notes and crystal-clear diagrams. We've also included practice questions to make sure you've mastered all the key points as you go along.

What's more, there's plenty of top advice for the exams and assessments. It's the best system for producing a top grade!

CGP — still the best! ☺

Our sole aim here at CGP is to produce the highest quality books — carefully written, immaculately presented and dangerously close to being funny.

Then we work our socks off to get them out to you — at the cheapest possible prices.

Contents

Section Five — Wood, Metals and Polymers

Section Six — Textiles

Section Seven — Electronic and Mechanical Systems

Section Eight — Designing and Making

Published by CGP

Editors:
Zoe Fenwick, Christopher Lindle, Camilla Simson, Ethan Starmer-Jones.

Contributors:
Ryan Ball, Sue Carrington, Juliet Gibson, Stephen Guinness, Brian Kerrush, John Nichols, Anthony Wilcock.

Consultant:
Ryan Ball

Proofreaders:
Anne Ainsworth, Chris McGarry, Glenn Rogers and Frances Rooney.

With thanks to Ana Pungartnik for the copyright research.

Photo of dressmaker on page 106: Blend Images/Shutterstock.com.

ISBN: 978 1 78294 752 3

Printed by Elanders Ltd, Newcastle upon Tyne.
Clipart from Corel®

Assessment Advice

Unlike most subjects, in Design and Technology you actually get to <u>design</u> and <u>make</u> something <u>useful</u>.

The Non-Exam Assessment is Worth 50% of your GCSE

1) The '<u>non-exam assessment</u>' (or <u>NEA</u> for short) is a <u>design and make</u> task — this involves making a <u>working prototype</u> and a <u>design portfolio</u>.

2) At the end of the first year of your course, you'll be given a choice of '<u>contextual challenges</u>'. These are <u>general topic areas</u>, e.g. 'in the workplace'. You'll <u>choose one</u> of these to <u>investigate</u> in your NEA.

3) Your teacher will give you as much help as they're allowed to by the exam board, so do <u>ask them</u>... but mostly it's <u>up to you</u> to make a <u>good job</u> of your NEA. You can dip into this book for extra help.

4) <u>Section 8</u> is focused on the <u>design and make</u> stuff, so it's a good place to <u>start</u>.

5) The <u>rest</u> of the book contains <u>useful information</u> for your NEA too though. If you're wondering about a particular <u>detail</u>, e.g. say, the properties of different woods, it's probably quickest to look that up in the <u>index</u> and go straight to those pages.

Only Put Relevant Stuff in your Portfolio

1) Your <u>teacher</u> will give you plenty of <u>guidance</u> about what to put into your <u>portfolio</u>. This is <u>documentation</u> of your work, which can be either <u>printed</u> or <u>digital</u>.

2) Your portfolio should lay out <u>each stage</u> of the <u>designing</u> and <u>making</u> of your prototype, including any <u>decisions</u> that you make as you go along. For example, it should include:

- A concise <u>summary</u> of <u>research</u> into your <u>contextual challenge</u>, including the <u>needs</u> and <u>wants</u> of the intended users, and a <u>design brief</u>.

- A <u>design specification</u> (see p.99), including how your <u>research</u> findings <u>helped</u> you to <u>decide</u> on the <u>design criteria</u>.

- <u>Documentation</u> (including photos) of any <u>models</u> you make. Don't just include the ones that worked — the ones that <u>didn't quite work</u> are useful, as you can explain <u>what was wrong</u> and how you fixed it.

- A <u>manufacturing specification</u> (see p.112) that explains the different stages of the making process to show <u>how you constructed</u> your prototype.

- An <u>evaluation</u> of your prototype, including <u>testing</u> it against what you originally set out to create in your <u>design brief</u>.

3) It's recommended that your portfolio is <u>20 pages</u> of <u>A3</u> (or the equivalent in A4) — you'll need to be <u>thorough</u> but write <u>concisely</u> to stick to this.

4) <u>Check</u> that you've used the right <u>technical words</u> and <u>spelled</u> things correctly. And make sure you've <u>explained things clearly</u> — get someone who <u>knows nothing</u> about your project to read it and see if it <u>makes sense</u>.

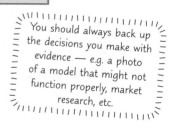
You should always back up the decisions you make with evidence — e.g. a photo of a model that might not function properly, market research, etc.

The Exam is Worth 50%

1) In the exam you could be tested on <u>anything</u> you've learned during the course — materials, tools, how to design things, how to make things, health and safety, environmental issues...

2) This book can help you to <u>learn</u> this stuff — and there are <u>questions</u> for you to <u>check</u> you've taken it in.

3) There are <u>answers</u> to these questions at the back of the book. There's a <u>glossary</u> at the back too, in case you need to sort out your mass production from your continuous production.

4) The <u>exam technique</u> section (pages 120-121) has some <u>worked examples</u> of exam-style questions, and some hints on how to make sure you get <u>top marks</u>.

Non-Exam Assessment — nope, it's not funny...

In your NEA, 60% of the marks are for designing and making your prototype. The rest of the marks are for investigating design ideas, your design brief and specification, and the analysis and evaluation of your work.

Technology in Manufacturing

So here we go, our journey in Design and Technology <u>starts here</u> with some key ideas about <u>manufacturing</u>...

Manufacturing is a System

1) Industry involves <u>manufacturing products</u>. Manufacturing involves <u>several processes</u> that are carried out to create a <u>finished product</u> from some <u>raw materials</u> (e.g. metal, textile, paper).

2) The process of manufacturing is a <u>system</u>. Products are made as a result of a number of different <u>stages</u> taking place one after the other.

Manufacturing on a large scale often involves machinery.

3) These systems can be split up into <u>three parts</u>:

INPUT	⟶	**PROCESS**	⟶	**OUTPUT**
This is all of the <u>materials</u>, <u>tools</u> and <u>equipment</u> that you start off with.		This is <u>what happens</u> to the input to change it into an output, e.g. measuring, cutting and forming.		The output is the <u>result</u> of the system — in other words, the <u>finished product</u>.

Advances in Technology can Improve Manufacturing

Technology <u>never stands still</u>. Advances can help manufacturing processes to run more <u>efficiently</u> — in recent years this has been enabled particularly by advances in <u>information and communications technology</u>.

Automation of manufacturing processes

1) Automation is the use of <u>machines</u> to do a task <u>automatically</u> (surprise, surprise) without much, or any, <u>human input</u>.

2) Automation has been a <u>key development</u> in manufacturing and is present in many <u>modern factories</u>. It can be used to carry out a <u>particular process</u> or manufacture a <u>whole product</u>.

E.g. in the car industry <u>robots</u> are used to do certain processes like <u>welding</u> parts together and <u>stamping</u> out metal body panels. These robots are <u>programmed by humans</u> but then run automatically. Humans are only needed to <u>monitor</u> the robots and <u>repair</u> them if they break down.

Advantages

- Robots can <u>increase</u> the <u>speed</u> of production as they can work <u>faster</u> than humans and don't need to <u>rest</u>. This also means robots can be <u>cheaper</u> to use than human workers.
- Robots can work with <u>high accuracy</u> as they perform the same task <u>consistently</u>, whereas humans can make <u>mistakes</u> (which have costs, e.g. time and materials). Robots can therefore increase the <u>quality</u> of manufactured products and <u>reduce costs</u>.
- Robots can be used in <u>dangerous situations</u> where it would be <u>unsafe</u> for humans.

Disadvantages

- Robots can <u>replace human workers</u> so there are <u>fewer jobs</u> for people to do.
- Robots can be <u>very expensive</u> to buy.
- Robots can't carry out tasks that require <u>human judgement</u>.

3) <u>Robots</u> are continually being developed to be <u>more intelligent</u> and to allow them to work <u>more closely</u> with <u>humans</u>. For example, robots exist that can be <u>easily programmed</u> by humans and can 'learn' how to do a task after being shown it by a human worker — this worker can then supervise the robot.

4) It's hoped that in the future factory robots will be developed that can deliver <u>consistent</u> and <u>accurate</u> work (as they do now) whilst also having some of the characteristics that only <u>human workers</u> currently offer (e.g. the ability to <u>make judgements</u> and come up with <u>ideas</u> for problem solving).

Smart technology in manufacturing

1) There are many examples of <u>computerised</u> machines that have been designed to carry out advanced tasks <u>accurately</u> (see CAD/CAM on page 4). <u>Smart technology</u> develops these machines further by <u>connecting them</u> with <u>other machines</u> and <u>sensors</u> in the factory so that they can <u>share data</u>.

2) These 'smart' machines can use this data to help <u>organise what tasks they need to do</u> <u>without</u> any <u>human input</u>. Smart machines can also receive <u>data</u> about <u>stock levels</u> and work out when they'll run out of materials and components (see next page).

3) Smart machines can be part of the '<u>Internet of Things</u>' — this describes the <u>connection</u> of loads of different machines, devices and other bits of technology to the <u>internet</u>. For machines this could involve <u>receiving</u> and <u>processing</u> online customer orders.

4) This use of smart technology can make automated manufacturing extremely <u>efficient</u>.

Technology in Manufacturing

Tracking of materials, tools, equipment and products

1) Factories contain <u>materials</u>, <u>tools</u>, <u>equipment</u> and the <u>products</u> themselves (both finished products and those that are partly made). Tracking of these items has been a <u>key development</u> in recent years.

2) This can be done automatically by <u>tagging</u> each box of material, tool or product etc. These tags can then be <u>scanned or detected</u> as the item moves around the factory and the status of the item kept <u>up to date</u> in a computer system.

3) Tracking the location of items has <u>many uses</u>. This includes <u>monitoring</u> the <u>stock levels</u> of raw materials and components, the <u>movement of products</u> around the factory as they're being made and the number of <u>finished products in stock</u>.

Tags like these are detected by wireless signals.

Communication systems

1) In modern automated factories, information can be passed around by <u>smart machines</u> so communication requires <u>no human input</u>.

2) <u>Human workers</u> can use <u>telephone</u>, <u>e-mail</u>, <u>video-conferencing</u> and other internet-based communication systems to share information.

3) Workers can also use <u>devices</u> to receive information. E.g. <u>warehouse</u> workers can use a <u>device</u> (like a tablet) to <u>direct</u> them around the warehouse in the <u>most efficient route</u> to collect items in customer orders.

Specialised buildings

1) New technology means that factories are being designed to <u>incorporate smart technology</u> and also to <u>minimise</u> their <u>environmental impact</u> (e.g. using new water recycling systems to reduce water use).

2) Factories can also be made of <u>modular components</u> that can be <u>taken apart</u>, <u>moved</u> or <u>added to</u>. E.g. <u>extra production space</u> can be added if needed.

3) Technological advances can also lead to a <u>reduced need</u> for <u>room</u>. E.g. the increased use of robots and automation means that factories can take up <u>less space</u>. <u>3D printing</u> (see p.68) can use a small machine, so manufacturing can <u>move out of large factories</u> to smaller, local spaces.

Different Ways of Manufacturing can help Maximise Efficiency

Flexible Manufacturing Systems (FMS)

1) <u>Flexible manufacturing systems</u> (FMS) consist of a set of <u>different machines</u> which carry out the different stages of production. These computer-controlled <u>automated</u> systems transport materials through the different processes and eventually <u>store</u> the finished product.

2) They're called <u>flexible</u> manufacturing systems because they are <u>easy to adapt</u> — it's fairly <u>quick</u> and <u>straightforward</u> to change the system if the <u>design</u> of the product being manufactured is <u>altered</u>. E.g. the <u>layout</u> of the system can be changed and <u>new machines</u> can be added in.

3) These systems are also easy to alter if the <u>level of production changes</u>, e.g. <u>extra machines</u> can be added to the system to <u>boost production</u>.

JIT DELIVERIES

Lean Manufacturing

1) Lean manufacturing comes from a Japanese approach to manufacturing that aims to minimise the <u>amount of resources used</u> and <u>waste</u> produced.

2) The aim of lean manufacturing is to <u>minimise costs</u> and <u>maximise efficiency</u>.

3) The <u>Just-in-Time (JIT)</u> system of stock control is a form of lean manufacturing. Materials and components are delivered <u>as they're needed</u> and used <u>as soon as they're delivered</u>.

Waste doesn't just refer to materials — it includes things like unnecessary use of time, transportation and movement of people, etc.

Advantages of JIT

- It <u>reduces</u> the amount of <u>space</u> needed for <u>storage</u> of materials and finished products (which <u>saves money</u> because you don't need to rent huge warehouses).
- It means there's less money tied up in materials that <u>aren't being used</u>.

Disadvantages of JIT

The system relies on materials and components being <u>delivered on time</u> and being <u>fault free</u> (there isn't time to return faulty goods). If these two things don't happen <u>money can be lost</u>.

These ways of manufacturing are becoming <u>more efficient all the time</u> as technology gets more advanced.

Revise Just-in-Time — better than two weeks after the exam...

Q1   Explain what is meant by a Just-in-Time system.   [2 marks]

Production Systems — CAD/CAM

CAD and CAM are a really important part of designing products and manufacturing them. They're used in loads of different industries from food packaging to component manufacture. Now to explain what they are...

CAD is Designing Using a Computer...

1) CAD stands for computer aided design.

2) It involves designing products on a computer, rather than using a pencil and paper.

3) CAD software packages allow you to make 2D or 3D designs. Examples of CAD software include Vectorworks®, SolidWorks®, TechSoft 2D Design® and ArtCAM®.

4) CAD helps designers model and change their designs quickly. It's easy to experiment with alternative colours and forms and you can often spot problems before making anything.

5) In 3D programs, you can view the product from all angles.

...and CAM is Making Using a Computer

1) CAM stands for computer aided manufacture.

2) It's the process of manufacturing products with the help of computers.

3) CAD software works out the coordinates of each point on the drawing. These are called x,y,z coordinates — x is the left/right position, y is forwards/backwards and z is up/down. The point where x, y and z meet is (0,0,0) — the datum.

4) CAM machines are computer numerically controlled (CNC) — they can follow the x,y,z coordinates and move the tools to cut out or build up your design.

5) There are a variety of different CAM machines available. They involve the subtraction (also called wastage) or addition of material.

6) Subtraction is where material is removed from a solid block to form an object.

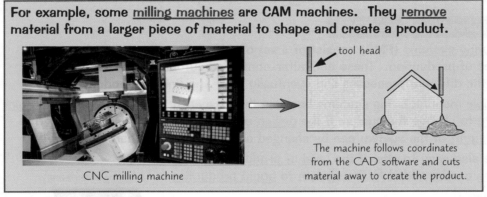

For example, some milling machines are CAM machines. They remove material from a larger piece of material to shape and create a product.

tool head

CNC milling machine

The machine follows coordinates from the CAD software and cuts material away to create the product.

Not an example of a CAM machine.

7) Addition is where material is added to build up an object.

For example, 3D printers are CAM machines. They add material (e.g. polymer) to build up an object and create a product.

Layers of material are built up to create the shape of the product using coordinates from the CAD software.

3D printing is increasingly popular in industry and this looks set to continue into the future — it allows production of items closer to end users. It's also become increasingly popular with DIYers and small producers making their own custom items using 3D printers.

8) Other examples of CAM machines are CNC routers and laser cutters (see next page).

Learn this CAD/CAM stuff or you'll look like a right tool...

CAD and CAM are all the rage in designing products these days. 3D printers have been around a while but have become more widely available recently. The thought of 3D printing a custom dog bowl on your kitchen table or in your technology lab would've sounded like witchcraft not so many years ago but now it's possible. Amazing.

Production Systems — CAD/CAM

CAM Machines Can be Used on Different Materials

1) There are CAM machines out there for all kinds of jobs — fear not...
2) Some CAM machines are 2-axis — they only use x and y coordinates so can only cut out 2D shapes.
3) Others are 3-axis machines — these use x, y, and z coordinates so they can cut out 3D shapes.

> CNC routers are able to cut out either 2D or 3D shapes from a block of material using different sized cutting tools — they're either 2-axis or 3-axis machines. They can also be used to engrave things. CNC routers can be used on plastics, wood and metals.

> Laser cutters are used to cut things — they can be used on plastic, wood, cardboard, fabrics and some metals. Laser cutters on high power settings cut right through the material. On lower power settings they can be used for engraving. Laser cutters can only cut through sheet materials — they're 2 axis machines so they can't cut out 3D objects. Die cutters (see p.53) are also used to cut out shapes from sheet materials.

> 3D printers can be used for rapid prototyping — they convert your design from an image on screen into a 3D model. They can be used to print with several different materials including plastic and wax (see p.68). It's also increasingly being used to manufacture final products.

There's more about prototyping on p.114-115.

CAD/CAM is Good for Global Companies

1) Nowadays it's easy to communicate with people all over the world via phones and the internet — as a result businesses can operate across the world. It's also easy to transport goods around the globe.
2) These factors combined with the development of CAD/CAM has affected manufacturing in two main ways:

Savings on shipping costs

Designs for products (e.g. cars) can be sent electronically...

...to be manufactured at a factory on the other side of the world, near where they'll be sold.

This means companies can save on the costs of shipping the finished products.

Savings on labour costs

Some companies design their products in one country...

...have them manufactured in another (where labour is cheaper)...

...and then ship them back to wherever they'll be sold.

3) CAD/CAM can offer lots of benefits to companies — these include the benefits of automation covered on page 2. However, there are drawbacks too which include the fact that computers can be affected by viruses, software problems and file corruption, which could potentially slow down production.

Can we make it? YES WE CAD/CAM...

Q1	What does CAD stand for?	[1 mark]
Q2	a) What is meant by machines being 'computer numerically controlled'?	[1 mark]
	b) Give an example of a CNC machine.	[1 mark]
	c) Describe the difference between CAM machines that carry out subtraction and CAM machines that carry out addition of material.	[2 marks]
Q3	Give an example of a material that a 3D printer can print with.	[1 mark]
Q4	Describe how using CAD/CAM can save on shipping costs for a business.	[2 marks]

Product Sustainability

When you make a product there's some sort of <u>impact</u> on the <u>environment</u>. Recently there's been an increasing focus on <u>minimising</u> the size of this impact to help <u>protect our planet</u>.

Sustainable Products are Better for the Environment

1) Sustainability means not causing <u>permanent damage</u> to the environment and not using up <u>finite resources</u> (ones that'll run out eventually). How sustainable a product is depends on:

> What <u>materials</u> are used to make it, for example:
> * Do they come from <u>non-finite</u> resources (ones that can be replaced) or <u>finite</u> ones?
>   E.g. most plastics are made from crude oil which is a finite resource. <u>Finite resources are</u> <u>slowly running out</u> and some could be <u>used up within 100 years</u> if new stocks aren't found.
> * Are they <u>recyclable</u> or <u>biodegradable</u> (this means they'll break down over time) — or will the used product permanently take up space in landfill?

For more about the social and ethical factors related to materials see p.35.

> The <u>processes</u> used to make the product, for example:
> * Does the process need lots of <u>energy</u>?
> * Does it create lots of <u>waste</u> or <u>pollution</u>?

<u>Wood</u> is a <u>non-finite</u> resource because trees can be <u>replanted</u>.

<u>Oil</u> is a <u>finite</u> resource — it will eventually <u>run out</u>.

2) Sustainability also depends on the <u>design</u> itself — how <u>long-lasting</u> and <u>efficient</u> the product is.

3) As you'll see on the next few pages, <u>new products</u> can have a <u>positive</u> and <u>negative</u> impact on sustainability and the environment.

All Products have a Carbon Footprint

1) A <u>carbon footprint</u> is the amount of <u>greenhouse gases</u> (e.g. carbon dioxide and methane) released into the atmosphere by <u>making</u>, <u>using</u> and eventually <u>reusing</u>, <u>recycling</u> or <u>disposing</u> of something at the end of its lifetime.

2) When a new product is made <u>resources</u> and the <u>finished product</u> itself are <u>transported</u>. This mileage causes the release of <u>carbon dioxide</u> because generally <u>fossil fuels</u> are burned to provide the <u>energy</u> for this transport (which emits <u>carbon dioxide</u>). E.g. carbon dioxide is released when the <u>raw materials</u> the product is made from are <u>transported to the factory</u> and when the product is <u>distributed</u> (e.g. transported to shops), <u>transported to the user</u> and taken to its <u>final disposal location</u> at the end of its lifetime. This transportation contributes to the product's <u>carbon footprint</u>.

3) A product's carbon footprint also includes the <u>emissions</u> that occur <u>during its manufacture</u> and during processes involved in its eventual <u>reuse</u>, <u>recycling</u> or <u>disposal</u>.

4) The more <u>energy</u> that's needed for these processes, the <u>bigger</u> the product's carbon footprint.

5) Carbon dioxide is usually released when a product is <u>used</u> too — e.g. when you charge your smartphone the <u>power</u> being used <u>contributes</u> to the product's carbon footprint.

The manufacture, use and disposal of products can also produce other forms of pollution — see p.8.

> So, carbon footprints aren't only linked to new products when they're being made and distributed — they <u>carry on</u> as they're being <u>used</u>. Much of a product's carbon footprint will be generated during these <u>early stages</u> though. This is why producing loads of <u>new products</u> is considered <u>bad</u> for the environment, whereas using a product for a <u>long time</u> is better for the environment.

6) The <u>bigger</u> a product's carbon footprint is, the <u>larger</u> its contribution to <u>global warming</u> (the warming of the Earth due to <u>greenhouse gases</u> in the atmosphere).

No Product Lasts Forever though

1) So making <u>new products</u> releases <u>emissions</u> which aren't good for the environment. The best solution would be to <u>never</u> make another <u>new product</u>. However, this obviously <u>isn't practical</u> because every product eventually comes to the <u>end of its usable life</u> — even a 'bag for life' that can be re-used many times will eventually <u>wear out</u>.

2) When a product wears out it becomes <u>waste</u> and has to be <u>disposed of</u> — this can cause <u>pollution</u>. Usually it's <u>replaced with a new one</u> too which also causes pollution... sigh.

Product Sustainability

3) Because of these things, it's important to make products using material that <u>has been</u> or <u>can be recycled</u>, or that'll <u>break down</u> when sent to landfill. For example:

- Many materials are <u>recyclable</u>. Products often carry <u>symbols</u> to tell you if they're recyclable. <u>Paper</u>, <u>glass</u> and <u>most metals</u> are all <u>easy to recycle</u>. Plastics can be recycled but it can be <u>expensive</u> and they <u>can't be recycled over and over again</u> (as their structure can <u>break down</u>). For <u>safety reasons</u>, this means that some items <u>can't be made from recycled plastics</u>, e.g. medical supplies.
- Some materials are <u>biodegradable</u> — they will <u>rot away</u> naturally (in a compost heap, say). <u>Wood</u> and <u>paper</u> are biodegradable and so are <u>some new plastics</u> (see next page).

4) Products made from <u>lots of components</u> and several <u>different materials</u> are awkward to recycle, because it's difficult to separate all the parts before you can recycle them.

5) However, '<u>Design for Disassembly</u>' tries to help with this problem. This is when a new product is designed so it can be <u>easily taken apart</u> at the <u>end</u> of its <u>lifetime</u> — this allows the <u>parts and materials</u> to be <u>reused or recycled</u> to make new products.

6) There's a <u>big drive at the moment</u> for all electronic and electrical goods to be made this way. It's an example of new products being designed to have a <u>reduced environmental impact</u> — there's more about this on the next page.

Some Products Aren't Designed to Last but Some Are

1) Some products are designed to <u>become obsolete</u> (<u>useless</u>) quickly. E.g. a disposable razor becomes blunt after a few uses and its blade can't be sharpened or changed. This is <u>planned obsolescence</u>.

2) Products with <u>up-to-the-minute</u> designs or technology become obsolete quickly because they <u>go out of fashion</u>. E.g. people often replace their <u>mobile phone</u> when a new, fancier model comes out.

3) Built-in obsolescence is generally <u>bad</u> for the environment — because more <u>materials</u> and <u>energy</u> are used to make <u>replacement</u> products.

4) However, products <u>can</u> be <u>designed to last</u> and have less of an impact on the environment. This involves making the product <u>durable</u>, and designing it so that parts can be <u>maintained</u> and <u>repaired</u> or <u>replaced</u>. This is known as <u>design for maintenance</u>.

- Most <u>household appliances</u> are designed to be <u>maintained</u> and <u>repaired</u> — this makes sense as they're <u>expensive</u> and just replacing them when they break would be <u>wasteful</u> as they are made of lots of parts.
- For example, washing machines can be maintained (e.g. cleaning filters) and they can be <u>repaired</u> by trained technicians when they break, e.g. by installing a <u>new part</u> rather than having to replace the whole thing.

- Another example is the idea of making <u>modular electronics</u>. This has been proposed for <u>mobile phones</u> and other electronics that are often updated like <u>tablets</u>, <u>laptops</u>, <u>cameras</u> and <u>music players</u>.
- The idea is that electronics are made up of different parts (or modules) that are designed so that they can be <u>individually upgraded and replaced</u>. E.g. with a modular phone you could <u>replace</u> the <u>processor</u> for a <u>faster one</u> when it starts to get a bit slow or the <u>battery</u> when it wears out. This is an alternative to getting a <u>completely new phone</u> when only one part of it breaks down, which <u>cuts down</u> a lot of <u>electronic waste</u> (which often ends up going to landfill because it's difficult to recycle).

Good news, this book is reusable HINT HINT...

This book is both reusable and recyclable. Best focus on the reusing bit before the exam though I reckon...

Q1	Define the term 'sustainability'.	[2 marks]
Q2	Describe what the carbon footprint of a product is.	[2 marks]
Q3	What is meant by 'Design for Disassembly'?	[1 mark]

Product Sustainability and Social Issues

Just a bit more on the environmental impacts of new products and then onto the social impacts. Woo...

Continuous Improvement of Products can be Bad and Good

1) Continuous improvement is a process by which manufacturers are constantly trying to improve their products, e.g. by changing the design of a product so it incorporates new technology (see page 10).

2) The continuous improvement of products can be environmentally damaging for many reasons, including:

- It can encourage consumers to replace existing products with new ones, which can lead to older models being disposed of. New, replacement products that are made have a carbon footprint (see p.6).
- New products being manufactured, packaged, transported and eventually disposed of can result in the increased usage of finite resources and environmental damage linked to the collection of these resources (see p.47).

3) However, manufacturers can also use the process of continuous improvement to make changes to their products that mean they have a reduced environmental impact. For example, new electronic products can be made using more efficient components than older models, therefore having lower carbon footprints (meaning they contribute less to global warming).

Many appliances have energy efficiency ratings, e.g. an A-rated fridge is more efficient than a D-rated one. Generally newer products have more efficient ratings than older ones and this can represent big power savings when they're being used.

4) Improvements that have a positive impact on the environment can also relate to other aspects of products and their manufacture, including:

- Improving the energy efficiency of manufacturing processes used to make products.
- Avoiding use of environmentally damaging chemicals (e.g. CFCs which contribute to global warming).
- Minimising the amount of material used in products where possible. This means fewer resources are consumed and products are lighter (so less energy is needed to transport them).
- Avoiding use of materials that are produced from unsustainable sources — e.g. timber sourced from rainforests.
- Using more renewable and/or biodegradable materials — e.g. bioplastics can be made that are made of (renewable) plant materials that also biodegrade (rot away) fully.
- Making products from materials that can be recycled and that are designed for disassembly (see previous page). This means that fewer new resources are needed, and often less energy is used — e.g. recycling old food cans takes much less energy than mining and processing new metal.
- Making products that are designed to last (see previous page).

You can Carry Out a Life Cycle Assessment

You can work out the potential environmental impact of a product by doing a life cycle assessment (LCA). These look at each stage of the life of a product — from the raw materials to when it's disposed of.

Choice of material

1) Hardwoods are often obtained from natural rainforests. Felling the trees destroys the habitat of pretty much everything living there (including people). Softwoods are a greener choice. They're usually from managed plantations — so more trees are planted and grow quickly to replace them. Recycled wood is also a good choice for the environment.

2) Metals have to be mined and extracted from their ores. Most plastics are made using crude oil, which is a finite resource so can't be replaced. These processes need a lot of energy and cause a lot of pollution.

Some products can be recycled — the materials can be used again in new products.

Manufacture

1) Manufacturing products uses a lot of energy and other resources. It can also cause a lot of pollution (e.g. some processes produce toxic gases).

2) You also need to think about waste material and how to dispose of it.

Product disposal

1) Products are often disposed of in a landfill site at the end of their life.

2) This takes up space and pollutes land and water (e.g. when paint washes off a product and gets into rivers).

Using the product

Using the product can also damage the environment. E.g. electrical products use electricity generated by burning fossil fuels, and paint can give off toxic fumes.

Product Sustainability and Social Issues

You Need to Know the 6 Rs

You can use the 6 Rs when designing to help reduce the impact that new products have on the environment and make the whole process more sustainable. There are things for consumers to think about too.

1) REPAIR — It's better to fix things instead of throwing them away. E.g. repairing a mobile phone can sometimes cost the same as buying a new one, but is better for the environment. Manufacturers can still make a profit by selling replacement parts.

2) RE-USE — Customers can extend a product's life by passing it on or using it again. Some people reuse products for other purposes, e.g. using an old car tyre to make a swing.

3) RECYCLE — Recycling uses less energy than obtaining new materials, e.g. by extracting metal. Products made from more than one material should ideally be easy to separate into recyclable stuff — clear 'recycle' labelling helps with this.

4) RETHINK — You should think about your design carefully — you might be able to make the product in a different way, e.g. a radio that you wind up instead of running off batteries.

5) REDUCE — Making long-lasting, durable products like rechargeable batteries reduces the number of products customers need to buy. It also means that manufacturers can cut down on energy use and transport.

6) REFUSE — You can refuse to buy a product if you think it's wasteful — e.g. if it has lots of unnecessary packaging, has travelled long distances round the world to the UK, or will be inefficient or costly to run.

Products have a Social Footprint

As well as having an environmental impact, the design and manufacture of products also impacts people.

Working conditions
- Firms have a moral and legal responsibility to provide safe working conditions for their employees.
- This is particularly relevant to people working in dangerous situations (e.g. agriculture and manufacturing). Keeping workers safe in these jobs can involve providing them with a safe working environment (e.g. factories designed to be as safe as possible), sufficient protective equipment (e.g. safety goggles) and training about how to work safely.
- There are initiatives for better working conditions. Companies that sign up to the Ethical Trading Initiative (ETI) agree to meet certain standards for working conditions, right across the supply chain — from the extraction of raw materials, through product manufacture, to the distribution of the products to the consumer.

Health impacts
- As you've seen on the past few pages, making products can release pollution (e.g. harmful chemicals). These can negatively affect the oceans and the atmosphere. For example, dioxins are made when fibres are bleached during the manufacture of paper, card and textiles. These chemicals are highly toxic to both animals and humans.
- Reducing levels of pollution is important because of the potential danger to human health that it can have. Efforts are underway to clean up manufacturing processes to reduce (or stop completely) the release of harmful pollutants in the environment. These include legal controls of certain chemicals and tough restrictions on what factories can release into the environment.

Safety first — always wear a hard hat when typing...

Q1 What is the purpose of a life cycle assessment? [1 mark]

Q2 Which of the following is **not** one of the 6 Rs?
  A: Repair **B**: Reassess **C**: Rethink **D**: Recycle [1 mark]

Q3 Give two ways that firms can help keep their employees safe. [2 marks]

Products in Society

New technology can impact <u>enterprise</u>, <u>people</u>, <u>culture</u> and <u>society</u> as a whole. <u>Read on</u> to find out how...

Entrepreneurs Take Advantage of Business Opportunities

1) <u>Enterprise</u> involves identifying new business opportunities, and then taking advantage of them. There's always a <u>risk of failure</u>, but the <u>reward</u> for a successful enterprise activity is <u>profit</u>.

2) Enterprise can involve <u>starting up</u> a <u>new business</u>, or helping an <u>existing one</u> to <u>expand</u> by coming up with new ideas (e.g. a new product). An <u>entrepreneur</u> is someone that does this.

3) When an entrepreneur or a business come up with something new this is called <u>innovation</u>. There are some <u>innovations</u> in business you need to know about:

Crowdfunding

- To start a new business or just launch an idea entrepreneurs need to <u>raise money</u>. Traditionally you could do this by applying for a <u>loan from a bank</u> or finding an <u>investor</u> (think Dragon's Den), but technology has provided another way called <u>crowdfunding</u>.

- Crowdfunding involves using websites (e.g. Kickstarter) to promote an idea to a huge number of people. People can then <u>choose to invest</u> an amount of money to help fund the idea — these people are called <u>backers</u>. Backers often get a <u>reward</u> for their investment — this could be a <u>free gift</u>, a <u>discount</u> on the product being funded or perhaps a <u>percentage share</u> of the new company once it has been funded.

Virtual marketing and virtual retail

- <u>Marketing</u> is the promotion of a product or service (e.g. magazine adverts). <u>Retail</u> is selling products and services to customers (e.g. selling products in high street shops). On the internet it's possible to do both of these things <u>virtually</u>.

- <u>Virtual marketing</u> includes promoting a product/service on social media (e.g. Facebook®), via email, and moving a product/service nearer to the top of the page in search engine results.

- <u>Virtual retail</u> is all about selling products and services on the internet. Many businesses sell their products <u>online</u> already but in the future it may become a <u>more realistic</u> experience with <u>virtual reality shops</u> that you can move around in from the comfort of your own living room...

Co-operatives and Fairtrade

- A co-operative is a type of business that is <u>owned and run</u> by its <u>members</u>. The members make the decisions about how the business is run and <u>profits are shared out</u> between them. Co-operatives include <u>small enterprises</u> like local shops and <u>large businesses</u> (e.g. The Co-operative Group).

- Groups of <u>farmers</u> often form co-operatives. This is when farmers work together so that they have stronger <u>negotiating power</u> with trading partners and can get <u>higher prices</u> for their produce. Co-operatives are a feature of <u>Fairtrade</u> which aims to ensure farmers are paid a fair price — see p.103.

Manufacturers Try to Continually Improve Products

1) Established product manufacturers are always looking for ways to make <u>more money</u>. One way to do this is by <u>improving</u> how they make their products, e.g. improving the <u>design</u> so that the product can be made more <u>easily</u>.

2) Manufacturers also redesign products in response to <u>market pull</u> or advances in <u>technology</u>...

Market Pull is About What Consumers Want

1) Designers often <u>design stuff</u> (and manufacturers make it) to satisfy the <u>wants</u> and <u>needs</u> of consumers — <u>consumer demand</u>.

2) Changing <u>fashions</u> and <u>social attitudes</u> affect the kind of products people want — consumer demand <u>won't</u> always be for the same <u>things</u> or <u>styles</u>.

For example, the car was invented to transport people from A to B, but now some consumers expect it to be more of a <u>status symbol</u>, demanding <u>luxury extras</u> like climate control, reversing cameras and fancy stereos.

Technology Push is About What Manufacturers Can Provide

1) In industry, research and development departments are always coming up with <u>new technologies</u>, <u>materials</u> and <u>manufacturing techniques</u>. This can drive the design of new products or <u>improve existing ones</u>.

2) Using new technology might make an existing product <u>cheaper</u>, <u>better</u> at its <u>function</u>, <u>more efficient</u> or <u>nicer-looking</u> — all things which will make products <u>more desirable</u>.

E.g. computers started off as <u>huge mechanical 'adding machines'</u>. Now, thanks to technologies including the microchip, they're <u>small</u> but really <u>fast and powerful</u>.

This process can have environmental impacts as it promotes increased consumption and waste — see p.8.

Products in Society

Technology Affects what Jobs People Do

1) Technology is always developing and this has an impact on the <u>jobs that are available</u> to do.

2) Increasing <u>automation</u> and the <u>use of machines</u> in manufacturing has <u>reduced</u> the number of people that are needed to carry out some <u>manual tasks</u> — this was covered earlier in the section (see p.2).

3) However, developments in technology have also opened up <u>new job opportunities</u>. For example, the development of <u>computers</u> and the <u>internet</u> has created lots of jobs. Workers are needed to <u>design</u> computers, <u>write programs</u> for them to run, design <u>computer games</u>, maintain <u>websites</u>, <u>teach</u> people how to use computers and devices... the list goes <u>on and on</u>.

4) Technology has also led to jobs <u>changing</u> as a result of the increased use of technology in the workplace.

> • Many <u>delivery drivers</u> use a device with <u>satellite navigation</u> to work out their <u>most efficient route</u> to complete their deliveries. This means they don't have to use a <u>map to navigate</u> or spend time working out the <u>best route</u> to do their deliveries.
> • Workers or manufacturing lines may <u>supervise a robot</u> rather than doing a technical task themselves.

Technology can Also Have an Impact on Culture

1) The <u>culture</u> of a particular <u>country</u> or <u>group of people</u> covers everything from their <u>religion</u>, <u>beliefs</u> and <u>laws</u> to their <u>language</u>, <u>food</u>, <u>dress</u>, <u>art</u> and <u>traditions</u>. Phew.

2) If you're designing a product aimed at a <u>specific</u> target market, you'll have to take into account the <u>views and feelings</u> of people from that particular culture — that way you don't <u>isolate or offend</u> anybody for <u>political</u>, <u>religious</u>, <u>gender</u> or <u>cultural</u> reasons.

3) New technology has the potential to <u>offend certain people</u>. For example, if a new material is developed that is made using <u>animal products</u> (e.g. animal fats), vegans and vegetarians may <u>choose not to buy</u> products made of the material.

4) New technology can also impact <u>fashion</u> and <u>trends</u>.

> • Fashion itself is continually affected by <u>new materials and techniques</u>. For example, in the 1960s when stretchy, body-hugging LYCRA® was introduced it revolutionised sportswear. New textiles and ways of making clothes will continue to give <u>new possibilities</u> for fashion.
> • Technology can also have an impact on <u>fashion trends</u>. The <u>internet</u> allows people to find out about fashion trends that are happening all over the world and new clothes can be seen by a <u>global audience</u> of people, e.g. via social media websites and blogs.

Products can Impact Particular Groups in Society

1) Products that are made can be <u>designed</u> to avoid having a <u>negative impact</u> on other people by being <u>sensitive to their needs</u>. Groups of people to consider include <u>disabled</u> and <u>elderly</u> people (see p.97).

2) Technology designed to suit <u>disabled and elderly users</u> includes <u>more accessible mobile phones</u> and <u>apps</u>. For example, mobile phones are sold with <u>large, easy-to-read buttons</u> that are easier to use for people with <u>less mobile hands</u> and/or <u>poor sight</u>. Apps that enable devices to be controlled entirely by <u>voice commands</u> have been developed to assist people with <u>physical disabilities</u>.

3) When designing products, people of different <u>religious groups</u> should also be considered.

4) If people believe that <u>religious symbols</u> have been misused or abused on a product this is likely to cause offence. For example, if symbols from a religion that <u>doesn't permit</u> alcohol consumption were used on the packaging of an <u>alcoholic drink</u> this is likely to be offensive.

If this page offends because it has too many words we're sorry...

... but you need to know this stuff, so get reading anyway. Hopefully you're not also offended by questions...

Q1   Tim is designing a mobile phone. Give two ways how technology push could affect his design.   [2 marks]

Q2   Why should people's views be considered when products are designed for a particular culture?   [1 mark]

Powering Systems

Energy is needed to manufacture products and to power systems. There are many ways of providing energy. You need to know about how they work, and the arguments for and against using different energy resources.

Fossil Fuels are Finite (Non-Renewable)

1) Fossil fuels are natural resources that form underground over millions of years. There are three main types — coal, oil and gas.

2) They're typically burnt to provide energy. For example, power stations burn fossil fuels to produce heat. This heats water to create steam, which drives a turbine. This turns a generator, generating electricity.

Boiler   Steam   Turbine   Generator

Fuel

Grid

For

1) Fossil fuels are reliable (at the moment) — there's enough to meet current demand.

2) Running costs of power plants and extraction costs for fossil fuels are quite low — so they're a fairly cost-effective way to produce energy.

Against

1) They're finite — they'll eventually run out.

2) The extraction of fossil fuels has social and environmental impacts (see p.47).

3) They release greenhouse gases when they are burned, which causes lots of environmental problems (see p.6).

Nuclear Power Produces Dangerous Waste

Nuclear power stations work in a similar way to fossil fuel power stations. The difference is that a process called nuclear fission produces the heat to create the steam that drives the turbines.

For

1) Nuclear power is reliable.

2) Nuclear fuel is quite cheap.

3) Nuclear power is clean — it produces very low levels of greenhouse gases.

Against

1) It uses fuel sources that are finite, e.g. uranium.

2) Nuclear power plants cost a lot to build, maintain, and decommission (shut down and make safe).

3) The waste produced is dangerous and difficult to dispose of.

4) Nuclear power always carries the risk of a major catastrophe, e.g. the Fukushima disaster in Japan.

Non-Finite (Renewable) Energy is Better for the Environment

Non-finite energy resources provide energy that won't run out — it can be renewed as it is used:

1) Wind power involves putting turbines in exposed places where they can take advantage of the wind. Each turbine has a generator inside it — the rotating blades turn the generator and produce electricity.

2) Solar cells generate electric currents directly from sunlight.

3) Tidal barrages are big dams built across river estuaries, with turbines in them. The flow of the tide spins the turbines and generates electricity.

4) Hydro-electrical power uses a dam to flood a valley. Water trapped in the top of the valley (the higher reservoir) is only allowed into the lower reservoir through turbines connected to generators.

5) Biomass involves burning waste wood or crop material to turn water into steam, which turns turbines, etc.

For

1) Renewable energy resources generally have a smaller environmental impact than other energy resources — once they've been set up, they provide clean energy.

2) After the initial set up costs, the energy provided is usually free.

Against

1) Initial set up costs are often quite high. They also generally don't produce as much energy as finite sources, so the time taken to pay back the initial expense is quite long.

2) Some types rely on external factors like the sun and wind, so they're less reliable.

3) Some people think renewable energy resources (e.g. wind turbines) spoil the landscape and look ugly. Wind farms can also be noisy, which can affect people living locally.

4) Some types of renewable energy resources can have a big impact on the environment, e.g. valleys are flooded for hydro-electric power, which can cause possible habitat loss.

Powering Systems

There is a Drive to use More Renewable Energy

Most of our electricity is produced using <u>fossil fuels</u> and <u>nuclear power</u>. Recently, there has been a <u>move towards</u> using more renewable energy resources. This has been triggered by <u>many things</u>...

- We now know that <u>extracting</u> and <u>burning</u> fossil fuels <u>damages</u> the <u>environment</u>.
- Many people think it's better to learn to <u>get by without</u> non-renewables <u>before</u> they <u>run out</u>.
- <u>Improved efficiency</u> in renewable power production mean it's becoming a <u>more attractive option</u>.
- <u>Pressure from other countries</u> and the <u>public</u> has meant that governments have begun to introduce <u>targets</u> for using <u>more renewable resources</u>, and for <u>cutting down</u> on <u>carbon dioxide</u> emissions (see p.6).

However, many people are <u>opposed</u> to the idea of renewables replacing fossil fuels, due to <u>factors</u> such as <u>initial cost</u>, <u>reliability</u>, the effect on the <u>landscape</u>, etc.

Energy Often Needs to be Stored

1) Most large power stations have to be kept running <u>all night</u> even though demand for electricity is <u>very low</u>. This means there's <u>extra</u> electricity being generated at night.

2) A <u>kinetic pumped storage system</u> is one of the best ways to <u>store</u> this extra energy for when it is <u>needed</u> during periods of <u>peak demand</u>.

3) The system operates using a <u>hydro-electric power station</u>. Here's how:

> - Rather than wasting it, 'spare' <u>night-time electricity</u> generated by <u>other power stations</u> is used to <u>pump water</u> in the <u>lower</u> hydro-electric reservoir <u>up</u> to the <u>higher</u> reservoir.
> - This water can then be allowed to <u>flow</u> through the <u>turbines</u> during periods of <u>peak demand</u> to <u>generate electricity</u> — this supplements the <u>steady delivery</u> from the other power stations.

Pumped storage uses the same idea as hydro-electric power but it isn't a way of generating power — it stores the energy that has already been produced by other power stations.

Batteries are Another Way of Storing Energy

1) Batteries <u>store chemical energy</u> — <u>chemical reactions</u> are used to generate an <u>electric voltage</u>, which pushes a <u>current</u> through a <u>circuit</u>.

2) They're often used to power <u>portable electronics</u>.

Electric currents, voltages and circuits are covered in more detail on p.24.

Alkaline batteries

1) Alkaline batteries are <u>normally disposable</u> — once they've gone <u>flat</u>, they <u>can't be used again</u>.

2) They're <u>recyclable</u>, <u>last a long time</u>, and <u>leak less</u> than some disposable batteries though, which <u>helps</u> to <u>reduce</u> their <u>environmental impact</u>.

3) Alkaline batteries have a <u>power output</u> that <u>gradually decreases</u> over time.

4) They're typically used for <u>toys</u>, <u>remote controls</u>, <u>torches</u> and <u>clocks</u>.

Rechargeable batteries

1) A rechargeable battery can be charged when it goes <u>flat</u> — electricity is used to <u>reverse</u> the <u>chemical reaction</u> that occurs whilst the battery is in use. This reforms the chemicals that are needed for the battery to <u>work again</u>.

2) They're <u>more expensive</u> than alkaline batteries, but they can be <u>used again and again</u> — this makes them <u>cheaper</u> in the <u>long run</u>.

3) They are more <u>environmentally-friendly</u> than <u>disposable batteries</u> because they can be used <u>more than once</u>.

4) Their <u>power output</u> remains <u>constant</u> until they run flat.

5) They're <u>built in</u> to some products (e.g. <u>mobile phones</u>, <u>laptops</u>, <u>electric cars</u>).

All this revision has drained my batteries...

Q1   List one advantage and one disadvantage of nuclear power.   [2 marks]

Properties of Materials

The <u>properties</u> of <u>materials</u> are their <u>characteristics</u> — all materials can be <u>described</u> in terms of their properties.

Different Materials Have Different Properties

1) The <u>properties</u> of a material determine what it's <u>useful for</u>. Thinking about these properties is super important when it comes to <u>designing</u> a <u>new product</u>.

2) Materials have <u>working properties</u>. Here are a few you'll need to <u>know</u> for the exam...

Strength

Strength is the ability to <u>withstand forces</u> without <u>breaking</u>. For example:

1) The rope in a tug-of-war resists <u>pulling</u> forces.
2) Bridge supports resist <u>compression</u> forces.
3) A surfboard resists forces trying to <u>bend</u> it.
4) <u>Fabrics</u> that contain <u>Kevlar</u>® fibres (see p.33) are really strong and <u>resistant to abrasion</u> — so they're used in <u>motorcycle clothing</u>.

Hardness

1) This is the ability to withstand <u>scratching</u>, <u>abrasion</u> or <u>denting</u>.
2) It's very important for tools that cut, like <u>files</u> and <u>drills</u>.

Toughness

1) If a material is <u>tough</u>, it is hard to <u>break</u> or <u>snap</u> — the material changes shape a bit instead.
2) <u>Armour</u> and <u>bulletproof vests</u> need to be tough.

Elasticity

1) Elastic materials can <u>stretch and bend</u> and <u>return</u> to their <u>original shape</u>.
2) A <u>spring</u> has good elasticity.

Malleability

1) Materials that are malleable can be <u>bent</u> and <u>shaped</u>.
2) Most <u>metals</u> are malleable — they can be hammered into <u>thin sheets</u> without breaking.

Ductility

<u>Ductile</u> materials can be <u>drawn</u> into a <u>wire</u>.

Copper is commonly used in electrical wires — it is both ductile and very electrically conductive.

3) Materials also have <u>physical properties</u>. For example:

Electrical conductivity

1) Electrical <u>conductors</u> let electricity <u>travel through them</u> easily. Electrical <u>insulators</u> don't.
2) <u>Electrical wires</u> need to be <u>conductors</u>, but the <u>coating</u> around the wires must be <u>insulating</u>.
3) <u>Metals</u> are good electrical <u>conductors</u>. <u>Plastics</u> tend to be good <u>insulators</u>.
4) Some <u>fabrics</u> are <u>blended</u> or <u>coated</u> with an electrically conductive material (e.g. a <u>metal</u>) — <u>gloves</u> can have <u>electrically conductive fingertips</u> so you can still operate <u>touchscreens</u> with them on.

Thermal conductivity

1) Thermal <u>conductors</u> let <u>heat travel</u> through them <u>easily</u>. Thermal <u>insulators</u> don't.
2) <u>Metals</u> are good thermal <u>conductors</u>. <u>Plastic</u>, <u>board</u> and <u>wood</u> are good thermal <u>insulators</u>.
3) <u>Pans</u> must be made from good thermal <u>conductors</u>, but their <u>handles</u> are often made from thermal <u>insulators</u>.

Some materials live in bungalows, some in semi-detached...

You need to be able to say why a material is suitable for a product, e.g. because it's tough. To do that, you really need to learn all of these properties. Try covering up the page and scribbling down as much as you can remember.

Properties of Materials

Fusibility

1) Materials with a high fusibility have low melting points — only a small amount of heating is required to convert these materials to liquids.

2) For example, solder has a high fusibility — this allows it to melt before the metals that are being soldered together (which have a lower fusibility).

Density

1) The density of a material is a measure of its mass per unit volume.

2) A table made of solid metal would likely be heavier to carry than an identical table made from plastic. This is because metals tend to be denser.

3) Density often has units of kg/m^3.

Absorbency

1) Fibres and fabrics that are absorbent are good at soaking up moisture. Paper towels are a good example.

2) Absorbent materials can be dyed easily, but they also dry slower and are vulnerable to stains.

3) Natural fibres (e.g. wool, cotton and cellulose fibres that make up paper) are absorbent.

4) Synthetic fibres (e.g. polyester and LYCRA®) are not absorbent.

You can also describe how resistant to moisture a material is. Materials that are poor at absorbing have a high resistance to moisture.

Non-Metals and Metals Have Different Physical Properties

Non-Metals

Non-metals tend to have a very different set of physical properties to metals.

Compared to metals, they are generally:

- More brittle (the opposite of tough)
- Not always solid at room temperature
- Poor electrical conductors
- Dull looking
- Less dense

Non-metals include all materials that don't contain metal — e.g. plastics, paper, textiles and wood.

Metals

Metals tend to have similar basic physical properties to one another. They're usually:

- Strong
- Malleable
- Good conductors of heat and electricity
- Not very fusible (have high melting and boiling points)

Metals and non-metals also have a wide range of chemical properties.

An Alloy is a Mixture

1) Alloys are a mixture of two or more metals, or a metal mixed with one or more elements.

2) The alloy is a new material — it has different properties to the individual metals it's made of.

3) Alloys are developed to have a specific set of properties. This is often done with future products in mind.

The Royal Albert Hall — good conductors always needed...

Q1 Describe what is meant by the following properties:

a) Fusible   b) Malleable   c) Ductile   [3 marks]

Q2 Jenny is designing a motorcycle jacket. She plans to make the jacket from a fabric.

Suggest a property that the fabric she chooses to use should have.   [1 mark]

Q3 Metal is used for radiators. Suggest one property that makes metal suitable for this use.   [1 mark]

Q4 Plastic is used for the casing of plugs. Suggest one property that makes plastic suitable for this.   [1 mark]

Q5 Absorbent materials soak up moisture. State two other properties of absorbent materials.   [2 marks]

Q6 Give two general physical properties of metals.   [2 marks]

Q7 Describe what is meant by an alloy.   [1 mark]

Paper, Board and Timber

Time to move on to materials made from <u>plants</u> and <u>trees</u> — that's right, it's <u>paper</u>, <u>board</u> and <u>timber</u>.

There are Various Types of Paper to Choose From...

1) Paper is pretty useful for <u>writing</u> and <u>sketching</u> (no, really).

2) There are lots of sorts of <u>paper</u> — each designed for a particular use.

3) <u>Cartridge paper</u> is high quality and has a <u>textured</u> surface — it's great for sketching with different drawing materials like pencils, crayons and inks.

4) <u>Layout paper</u> is <u>thin</u> and <u>translucent</u> (you can see light through it) and is used for <u>general design work</u> — particularly sketching ideas.

5) <u>Tracing paper</u> is <u>semi-transparent</u>, and is used to <u>copy images</u>.

6) <u>Grid paper</u> may have a <u>square</u> or <u>isometric pattern</u> printed on it. Square grid paper's useful for <u>orthographic and scale drawings</u> (see p.110-111) and isometric paper's good for <u>isometric drawings</u> (see p.108).

7) <u>Bleed-proof paper</u> is used by designers when drawing with <u>felt-tips</u> and <u>marker pens</u>. The ink doesn't spread out (<u>bleed</u>) — it stays put.

Square grid paper

Isometric grid paper — this paper also comes without lines (i.e. just dots where the lines cross).

Ink bleeds on some paper because the paper fibres suck the ink away.

...and a Selection of Boards too

1) The weight of paper and board is measured in <u>gsm</u> (grams per square metre). Above <u>200 gsm</u>, it's not paper any more — it's <u>board</u> (also known as <u>card</u> or <u>cardboard</u>).

2) Board is often used in packaging because of its <u>low cost</u> compared to other packaging materials, and its <u>high strength-to-weight ratio</u>.

> <u>Solid white board</u> has a high quality bleached surface, which is ideal for <u>printing</u>. It's used loads for <u>primary packaging</u> — packaging for individual items. (<u>Secondary packaging</u> is used to contain lots of the same item — see below.)

> <u>Ink jet card</u> is card used for <u>ink jet printing</u>. It's designed so that the ink <u>doesn't bleed</u> — this allows the printed image to be <u>sharply defined</u> and of a <u>high quality</u>.

> <u>Corrugated card</u> is made up of a <u>fluted inner core</u> sandwiched between <u>two outer layers</u> (the <u>liner</u>), which can be <u>printed</u> on. The flutes add <u>strength</u> and <u>rigidity</u> — this is useful in a lot of <u>secondary</u> packaging to <u>protect</u> products during transit.

> <u>Duplex board</u> has a <u>different colour</u> and <u>texture</u> on <u>each side</u>. It's often used where only <u>one surface</u> is <u>seen</u>, so that only one side needs to be <u>smooth</u> for <u>printing</u>. This board is often used for <u>food packaging</u>.

> <u>Foam core board</u> is made by sandwiching expanded <u>polystyrene foam</u> between 2 thin layers of <u>card</u>. It's <u>stiff</u>, <u>lightweight</u> and the thin outer card layer <u>can be scored</u> (see p.53). It's good for <u>making models</u> and <u>mounting posters</u>.

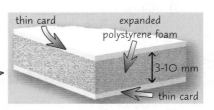

thin card

expanded polystyrene foam

3-10 mm

thin card

Foam core board and foil lined board are both laminated (see p.37) and composite materials (see p.33).

> <u>Foil-lined board</u> is a board with an <u>aluminium foil lining</u>. It's often used to <u>package food</u> (see p.48) — the foil keeps <u>flavours in</u>, and <u>air</u> and <u>moisture out</u>.

Mmm, paper, my favourite — just add fish and chips...

All these different types of paper and board are here for a reason — each one's best at a different job. It's a good idea to learn some of their properties in case you get asked which sort you should use in your exam *paper*...

Paper, Board and Timber

Wood is Either Softwood or Hardwood

Wood comes in two sorts — expensive hardwoods and cheaper softwoods.
Timber refers to sawn chunks of solid wood that are used as building materials.

Softwoods

Softwoods grow in colder climates and are fast growing — this makes
them fairly cheap and readily accessible. The trees have leaves like
needles, are usually evergreen and have cones (e.g. pine).

a knot

1) Pine is yellow with brown streaks. It's quite strong and
cheap — but knotty, which makes it harder to work with.
It's used for telegraph poles, fences and cheap furniture.

2) Larch has an attractive yellow to reddish-brown colour.
It's harder, tougher and more durable than most softwoods.
It's also resistant to rot, which makes it good for decking,
cladding the outside of buildings, and fence posts.

3) Spruce is a reddish brown colour. It's hard and has a
good strength-to-weight ratio — but it's also knotty
and not very durable. It's used for structural purposes
both outside and in — aircraft, crates and ship masts.

Hardwoods

Hardwoods usually grow in warm climates and are slow growing — so they're generally
more expensive than softwoods. The trees have broad, flat leaves and are usually
deciduous (they lose their leaves in autumn). The wood tends to have a tighter grain and
be denser and harder than softwood, although there are exceptions to this, e.g. balsa.

1) Oak is light brown. It's tough, durable and very strong. It also has attractive
grain markings, especially when quartersawn (see p.44), and finishes well, so
it's used in a lot of interior panelling, flooring and furniture.
However, it does corrode steel screws and fittings.

2) Mahogany is a red-brown colour. It's durable and easy to work with,
but it's expensive, so it's used for good quality furniture.

3) Beech is pinkish-brown. It's hard enough to resist being dented, and can
be bent using steam. It's used for chairs and toys.

4) Balsa is a white or tan colour. For a hardwood, it has a very low density and
is very soft. Its softness makes it easy to cut and shape — this combined
with a high strength-to-weight ratio, makes it great for modelling (see p.106).

5) Ash has a pale cream colour. It's tough and absorbs shock well, so it's
used for tool handles and wooden sports equipment (e.g. baseball bats).
It's also attractive so is used for furniture.

After one step Jim realised his balsa staircase had been a mistake...

Q1 What happens when ink bleeds? [1 mark]
Q2 What is the difference between paper and board? [1 mark]
Q3 Give a property of spruce that makes it suitable as a material for crates. [1 mark]
Q4 a) State one example of a hardwood. [1 mark]
    b) For the hardwood you gave in part a), give an example of how it is used. [1 mark]

Metals, Alloys and Polymers

Some metals are <u>pure</u> metals and others (alloys) are <u>mixtures</u> of a metal with another element.
Both types of metal can be classified into <u>two</u> basic groups — <u>ferrous</u> and <u>non-ferrous</u>.

Ferrous Metals Contain Iron

1) These metals or alloys are mostly made up of <u>iron</u>. Because of this, almost all of them are <u>magnetic</u>.

Metal	Properties	Uses
Cast Iron Iron + 2-3.5% Carbon + Silicon	Very <u>strong</u> if compressed, but <u>brittle</u> and <u>not malleable</u>.	Bench vices, car brake disks, manhole covers
Low-Carbon Steel (also known as Mild Steel) Iron + 0.05-0.35% Carbon	Quite <u>strong</u> and <u>cheap</u>, but <u>rusts</u> easily and <u>can't</u> be hardened.	Car bodies, screws, nuts, bolts, nails, washing machines
High-Carbon Steel (also known as Tool Steel) Iron + 0.5-1.5% Carbon	<u>Harder</u> than low-carbon steel, <u>can</u> be hardened. <u>Not</u> as <u>easy to work</u> though and it <u>rusts</u>.	Tools, e.g. chisels, files, saws, drills

Remember, malleable materials are bendy.

2) <u>Protective coatings</u> (e.g. paint, enamel) are often applied to <u>prevent</u> ferrous metals from <u>rusting</u>.
An exception to this is <u>stainless steel</u> (see below) — it's <u>resistant to rust</u> so doesn't need to be coated.

Non-Ferrous Metals Don't Contain Iron *(surprise surprise...)*

These metals or alloys <u>don't contain iron</u> so <u>don't rust</u> — useful if they're likely to be <u>exposed to moisture</u>.

Metal	Properties	Uses
Aluminium	<u>Lightweight</u> and <u>corrosion-resistant</u>, but <u>expensive</u> and <u>not</u> as <u>strong</u> as steel (although strong for its weight). <u>Hard to join</u>.	Aeroplanes, cans, ladders
Brass 65% Copper + 35% Zinc	<u>Quite strong</u>, <u>corrosion-resistant</u>, <u>malleable</u>, <u>ductile</u> and looks good (nice <u>golden bronze</u> colour).	Electrical parts, door handles, taps
Copper	<u>Relatively soft</u>, <u>malleable</u>, <u>ductile</u> and a <u>very good electrical conductor</u>.	Electrical wiring, pipes.
Tin	<u>Soft</u>, <u>corrosion-resistant</u>, <u>malleable</u> and <u>ductile</u>. Has a <u>low melting point</u>.	Foil, tin cans (tin-plated steel), alloying metal in solder (see p.71).
Zinc	<u>Not very strong</u>, <u>corrosion-resistant</u>.	Coating steel (e.g. on nails, buckets and watering cans).

Alloys can have Really Useful Properties

<u>Alloys</u> are formed when <u>one or more elements</u> are <u>combined</u> with a <u>metal</u> (see p.15). Here are some <u>examples</u> you need to know:

High Speed Steel

1) <u>High speed steel</u> contains <u>iron</u>, more than <u>0.6% carbon</u> and <u>other metals</u> including chromium, tungsten, and vanadium.
2) It keeps its <u>hardness</u> when <u>heated</u> to high temperatures, so it's used in <u>high speed cutting tools</u> that <u>get hot</u> when used.

Brass (copper + zinc)

1) Brass is <u>harder</u> and <u>stronger</u> than both copper and zinc.
2) It is <u>malleable</u>, <u>ductile</u> and a <u>good electrical conductor</u> (like copper), and is <u>resistant to corrosion</u> (like zinc).

Stainless steel (iron + carbon + chromium + nickel)

1) <u>Cast iron</u> and <u>mild steel</u> (iron + carbon) are strong but <u>rust (corrode) easily</u>.
2) Adding <u>chromium</u> and <u>nickel</u> increases <u>strength</u>, <u>toughness</u> and <u>ductility</u>, and <u>decreases rust</u>.
3) It is used in <u>surgical equipment</u>, <u>sinks</u> and <u>cutlery</u>.

Metals, Alloys and Polymers

Next up, plastics. Plastics are <u>made</u> of <u>polymers</u> (see p.45) but sometimes they just get <u>called</u> 'polymers'. There are <u>two main sorts</u> — <u>thermoforming</u> and <u>thermosetting plastics</u>.

Thermoforming Plastics are Recyclable and Bendy

1) Thermoforming plastics <u>don't resist heat</u> well — they're easily formed into different shapes by <u>heating</u>, <u>melting</u> and <u>remoulding</u>.

2) This means they're <u>easy to recycle</u> — they're <u>ground down</u>, <u>melted</u> and <u>re-used</u>.

Thermoforming plastic	Properties and Uses
Acrylic (PMMA)	<u>Hard</u>, <u>stiff</u> and <u>shiny</u>. <u>Resists weather</u> well. Can be used to make motorcycle helmet visors, baths, signs, etc. Quite <u>brittle</u> (not tough).
High-density polyethylene (HDPE)	<u>Stiff</u> and <u>strong</u> but <u>lightweight</u>. Used for things like washing-up bowls, baskets, folding chairs and gas and water pipes.
Polyethylene terephthalate (PET)	PET is a <u>polyester</u> that is <u>light</u>, <u>strong</u> and <u>tough</u>. Used to make see-through drink bottles and fibres for clothes.
High impact polystyrene (HIPS)	<u>Rigid</u> and <u>fairly cheap</u>. Used for vacuum forming and fabricating boxes like CD cases or smoke detector casings.
Polyvinyl chloride (PVC)	<u>Quite brittle</u>, <u>cheap</u> and <u>durable</u>. Used for blister packs, window frames, vinyl records and some clothing.
Polypropylene (PP)	Quite <u>tough</u> and <u>flexible</u>. Can be made in a wide variety of <u>bright colours</u>. Used for plastic chairs — its flexibility makes it comfortable.

Thermosetting Plastics are Non-Recyclable and Rigid

1) Thermosetting plastics <u>resist heat</u> and <u>fire</u> (so they're used for electrical fittings and pan handles).

2) <u>Thermosetting plastics</u> (unlike thermoforming plastics) undergo a <u>chemical change</u> when <u>heated</u> and <u>moulded</u> to make a product — they <u>permanently</u> become <u>hard</u> and <u>rigid</u>.

3) This means they're <u>non-recyclable</u> — they <u>can't</u> be melted and reshaped again.

Thermosetting Plastic	Properties and Uses
Epoxy Resin (ER)	<u>Rigid</u>, <u>durable</u>, <u>corrosion-resistant</u> and a <u>good electrical insulator</u>. Used for circuit boards and wind turbine rotor blades.
Urea-Formaldehyde (UF)	<u>Hard</u>, <u>brittle</u> and a <u>good electrical insulator</u>. Used for things like plug sockets and cupboard handles.
Melamine-Formaldehyde (MF)	<u>Strong</u> and <u>scratch-resistant</u>. Used to laminate chipboard and for plates and bowls.
Phenol-Formaldehyde (PF)	<u>Hard</u> and <u>heat-resistant</u>, it's very <u>easily moulded</u> into bottle caps, snooker balls, etc. Also mixed with other materials to form a <u>composite</u> (see p.33).
Polyester Resin (PR)	<u>Hard</u>, <u>stiff</u>, <u>cheap</u> and a <u>good electrical insulator</u>. Added to glass fibres to form glass-reinforced plastic (see p.33), e.g. for kayaks. Waterproof, so used in shower stalls and garden furniture.

Make sure you know the difference between thermoforming and thermosetting plastics, and can give examples of each.

Recycling — it's not biking back the way you came...

Thermosetting plastics can't be remoulded — once they're set, they are set permanently. Like when you pull a funny face and the wind changes. Something like that, anyway...

Q1 What is a non-ferrous metal? Give an example of one. [2 marks]

Q2 Explain why stainless steel is useful for making products that are used outdoors. [2 marks]

Q3 Give two differences between thermoforming and thermosetting plastics. [2 marks]

Q4 Give a property of melamine formaldehyde that makes it suitable for laminating worktops. [1 mark]

Textiles

Textiles, also called <u>fabrics</u>, can be made from <u>natural</u> or <u>synthetic fibres</u>.

Natural Fibres come from Natural Sources

1) Natural fibres can be <u>harvested</u> from <u>plants</u> and <u>animals</u> — <u>cotton</u> comes from <u>plants</u> and <u>wool</u> from <u>sheep</u>.

2) These are <u>renewable resources</u> (i.e. you can always produce more of them). They are also <u>biodegradable</u> (can be broken down by microorganisms in the ground when disposed of) and <u>recyclable</u> — natural fibres are therefore <u>fairly sustainable</u>.

3) In general, natural fibres are <u>absorbent</u> and <u>strong</u> when dry, but have <u>poor resistance</u> to <u>biological damage</u>, e.g. from moths and mould.

Fibre	Properties			Used in these fabrics	Uses
	Appearance	Good points	Bad Points		
Cotton	Smooth	Strong, hard-wearing, absorbent, comfortable to wear, feels quite cool in hot weather, easy to wash & add colour to, doesn't cause allergies, non-static, fairly cheap.	Creases easily, high flammability, poor elasticity, can shrink when washed, dries slowly.	E.g. denim, corduroy, calico	Jeans, T-shirts, blouses, soft furnishings.
Wool	Soft or coarse	Warm, absorbent, good elasticity, low flammability, crease-resistant, available in lots of fabric weights.	Can shrink when washed, dries slowly, can feel itchy, fairly expensive.	E.g. knitted fabrics, Harris Tweed, gabardine, jersey, felt	Suits, jumpers, coats, dresses, carpets.
Silk	Very smooth and glossy.	Smooth, resistant to shrinking/stretching, absorbent, good drape, low flammability, comfortable to wear, lightweight, sun-resistant.	Creases easily, might not wash well, weak when wet, expensive.	E.g. organza, chiffon, satin	Lingerie, underwear, dresses, shirts, ties.

Synthetic Fibres are Man-Made

1) Synthetic fibres are made from <u>polymers</u> (long chain molecules). These molecules come mainly from <u>coal</u> and <u>oil</u> — <u>non-renewable fossil fuels</u>. Synthetic fibres are therefore much <u>less sustainable</u> than natural fibres.

2) Synthetic fibres can be given many <u>different properties</u>. In general, they're <u>resistant</u> to <u>biological damage</u>, and can be <u>changed by heating</u> to form different shapes and textures. However, they're <u>not very absorbent</u>, so they can be hard to <u>dye</u>.

No matter what they threw at him, the new, synthetic James Bond just wouldn't dye.

Fibre	Properties			Used in these fabrics	Uses
	Appearance	Good points	Bad Points		
Elastane	Soft	Extremely elastic (stretches up to 7 times its length), strong, hard-wearing, lightweight, keeps its shape well, resists sun/biological damage.	Not absorbent, high flammability, not biodegradable.	E.g. LYCRA®	Sportswear, underwear, combined with other fibres to add stretch.
Polyester	Smooth, can have many different finishes.	Strong (even when wet), hard-wearing, low flammability, good elasticity, cheap, resists creasing, dries quickly, resists biological damage.	Not absorbent, not biodegradable, damaged by strong acids, melts as it burns (so harmful if clothes catch fire).	E.g. DACRON®	Sportswear, bed sheets, curtains, cushions, padding, tablecloths.
Polyamide	Can have many different finishes.	Strong, hard-wearing, warm, good elasticity, crease resistant, resists biological damage, fairly cheap.	Not very absorbent, damaged by sunlight, melts as it burns.	E.g. nylon	Sportswear, socks, tights, furnishings, carpets.

Textiles

Fabrics are Made with Fibres that are Spun into Yarns

Before we talk about fabrics, you need to know a bit more about <u>how they are made</u>...

1) <u>Yarns</u> are threads that are <u>woven</u> or <u>knitted</u> to make fabrics.
2) Yarns are made of <u>fibres</u> (tiny hairs). These come in either <u>short</u> lengths (<u>staple fibres</u>) or <u>longer</u> lengths (<u>filaments</u>).
3) <u>Filaments</u> can be <u>spun</u> (twisted together) or used <u>as they are</u>. <u>Staple</u> fibres are <u>spun</u> to produce <u>yarns</u>.

Yarns made from <u>filaments</u> are <u>smooth</u>...

...while yarns made from <u>staple fibres</u> tend to be <u>hairier</u>.

4) Yarns are available in different <u>thicknesses</u>. A <u>1-ply yarn</u> is a single yarn, a <u>2-ply yarn</u> is two yarns <u>twisted</u> together, a <u>3-ply yarn</u> is three yarns twisted together... and so on...

Fibres and Yarns are Used to Make Fabrics

There are <u>three main ways</u> of <u>turning fibres</u> or <u>yarns</u> into a <u>fabric</u> — they are <u>weaving</u>, <u>knitting</u> or <u>bonding</u>. There's more about each of these types of fabric on the next page.

① **WOVEN FABRICS** are made by <u>interlacing</u> (crossing alternately over and under) <u>two sets</u> of <u>yarns</u>.

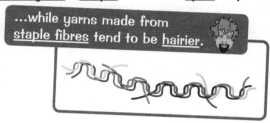

② **KNITTED FABRICS** are made by <u>interlocking one or more yarns</u> together using <u>loops</u>. The loops <u>trap air</u>, so they <u>insulate</u>. They <u>stretch</u> more than woven fabrics.

③ **NON-WOVEN FABRICS** are layers of fibres (not yarns) held together by <u>bonding</u> or <u>felting</u>. They <u>don't fray</u>, and can be <u>cut</u> in <u>any direction</u> — which means there's <u>little waste</u> when laying out patterns (see p.117). However, they <u>don't stretch</u> and <u>aren't</u> very <u>strong</u>.

Rapunzel, Rapunzel — let down a yarn of many staple lengths...

It's easy to get the properties of all these fabrics mixed up. Try to picture an example of something made from each one and then think about why it has been chosen for this specific purpose. Now then, time for some questions...

Q1 State two reasons why natural fibres are sustainable. [2 marks]
Q2 Give two examples of synthetic fibres. [2 marks]
Q3 Hazel is designing a T-shirt. She wants to make it from a fabric that isn't expensive and then dye it bright green.
   a) Given her requirements, what is a suitable material for Hazel to use? [1 mark]
   b) Explain your answer to a). [2 marks]
   c) Give one disadvantage of using this fabric in clothes. [1 mark]
Q4 Fibres come as staple lengths or filaments. Which of the two is longer? [1 mark]
Q5 How are woven fabrics made? [1 mark]

Textiles and Manufactured Boards

It's time to go into a bit more <u>detail</u> about how each type of fabric is <u>made</u>. Here we go...

Fabrics can be made from Woven Yarns...

1) Different <u>woven fabrics</u> are made by interlacing <u>two</u> sets of yarns — the <u>weft</u>, which travels from <u>right to left</u> (purple in the diagram below) and the <u>warp</u>, which travels <u>up and down</u> the weave (green in the diagram below).

- The <u>simplest</u> weave is the <u>plain weave</u> — the weft yarn passes over and under <u>alternate</u> warp yarns, making it <u>unpatterned</u>.
- It's <u>hard-wearing</u> — <u>strong</u> and holds its <u>shape</u> well. It has a <u>smooth</u> finish (making it good for printing on).
- It's the <u>cheapest</u> weave to produce and is used to make loads of fabrics, especially <u>cotton-based</u> ones.

The <u>edge</u> of a woven piece of fabric, where the weft yarns wrap around the warp yarns, is called the <u>selvedge</u>.

2) Fabrics are woven using <u>looms</u>. Smaller looms are <u>hand-operated</u> and used to create small pieces. However, <u>industrial looms</u> are used to <u>produce woven yarns on a large scale</u>. They are <u>computer operated</u> and much faster than weaving by hand, meaning that fabrics can be <u>mass produced</u> at <u>high speed</u>.

3) Woven fabrics are used for <u>shirts</u>, <u>upholstery</u> and <u>trousers</u> etc.

... Knitted Yarns...

There are a <u>two</u> main types of <u>knitted fabrics</u> — <u>weft-knitted</u> and <u>warp-knitted</u>:

Weft-knitted

1) The yarn runs <u>across</u> the fabric, making interlocking <u>loops</u> with the <u>row</u> of yarn <u>beneath</u>.
2) These fabrics <u>stretch</u> and can <u>lose their shape easily</u>.
3) If the yarn breaks it can <u>unravel</u> and make a 'ladder'.
4) Weft-knit fabrics can be produced by <u>hand or machine</u>.
5) They're used for <u>jumpers</u>, <u>socks</u> and <u>T-shirts</u>.

Warp-knitted

1) The yarns run "<u>up</u>" the fabric, in <u>loops</u>, which <u>interlock vertically</u>.
2) They're <u>stretchy</u> but still <u>keep their shape</u>.
3) These fabrics are <u>hard to unravel</u> and are less likely to ladder.
4) They're made by <u>machine</u> — and the machines can be expensive.
5) <u>Tights</u>, <u>swimwear</u>, <u>fleeces</u>, and some <u>bed sheets</u> are all made from warp-knitted fabrics.

... or Non-Woven Fibres

Bonded Fabrics

1) These are "<u>webs</u>" of synthetic fibres that are <u>glued</u>, <u>needle-punched</u>, <u>stitched</u>, or <u>melted</u> together.

Stitch-bonded non-woven fabric

fibres

stitching

2) They're used for <u>interfacing</u> (putting in between layers of fabric for thickness, e.g. in collars — see p.37), <u>artificial leathers</u> and <u>disposable cloths</u> (e.g. medical masks).

Felted Fabrics

1) <u>Felting</u> is an older way of making non-woven fabric. Felt is made by combining <u>pressure</u>, <u>moisture</u> and <u>heat</u> to <u>interlock</u> a <u>mat</u> of <u>wool fibres</u>.
2) Felt can be used for <u>carpet underlay</u>, <u>craft material</u>, <u>hats</u>, <u>jewellery</u> and <u>snooker table coverings</u>.

Textiles and Manufactured Boards

Fibres can be Combined by Blending or Mixing

Fabrics made from a <u>combination</u> of different <u>fibres</u> can give you <u>better</u> properties.

There are <u>two ways</u> of combining fibres to get fabrics with combined properties — <u>blending</u> fibres within yarns, and <u>mixing</u> different yarns (where each yarn is made from one type of fibre). Here's how:

1) A **BLEND** is when two or more <u>different fibres</u> are combined to produce a <u>yarn</u>.

2) This mixed yarn is then <u>woven</u> or <u>knitted</u> to make a <u>blended fabric</u>.

1) A **MIX** is when a fabric is made up of two or more <u>different types of yarn</u>.

2) The <u>two</u> different types of yarn can then be <u>knitted</u> or <u>woven</u> together to make a <u>mixed fabric</u>.

Combining <u>cotton</u> and <u>polyester</u> fibres is one of the most common <u>blended</u> fabrics.
The resulting fabric:
- is even <u>stronger</u> and remains <u>hard-wearing</u>
- is <u>less absorbent</u> — so dries more quickly
- is <u>soft</u> and <u>comfortable</u>
- resists <u>creasing</u> (is easier to iron)
- <u>doesn't shrink</u> easily
- BUT is <u>highly flammable</u>

Boards can be Manufactured

<u>Processed</u> pieces of <u>wood</u> can be combined with <u>glue</u> and <u>compressed</u> into <u>panels</u> — this forms new materials called "<u>manufactured boards</u>" or "<u>manufactured timbers</u>" (see p.44).
Here are a few examples you should know about:

Medium Density Fibreboard (MDF)

1) MDF is made from <u>tiny fibres</u> of <u>softwood timber</u> held together by <u>glue</u>.

2) As it's just bits of timber stuck together, it has <u>no natural grain</u>.

3) It's pretty <u>cheap</u>, <u>dense</u> and has a <u>smooth uniform surface</u> that <u>takes paint</u> and <u>other finishes</u> well. However, it's <u>porous</u> so can be <u>damaged by moisture</u>.

4) It's often used for <u>shelves</u> and <u>flat-pack furniture</u>.

Plywood

1) Plywood is made up of several <u>layers</u> of softwood or hardwood, glued together with their <u>grain</u> at <u>right angles</u> to one another (see p.37).

2) That structure makes it <u>very strong</u> for its weight and thickness, compared with solid wood.

3) It's a very popular manufactured board, used for <u>building</u> and <u>furniture</u>.

Chipboard

1) Chipboard is made by compressing <u>wood chips</u>, <u>shavings</u> and <u>sawdust</u> together with glue. It's usually used with a <u>veneered surface</u>.

2) It's <u>cheap</u> but <u>not very strong</u>. It's also <u>absorbent</u>, so can be easily <u>damaged by moisture</u>.

3) Chipboard is often used in <u>cheap self-assembly furniture</u>.

Mixed fibres — not like putting prunes on your bran flakes...

Mixing and blending fibres is really useful — you get a whole load of new properties to play with...

Q1	What are the names of the two sets of yarn used in weaving?	[2 marks]
Q2	Name the piece of equipment used for weaving.	[1 mark]
Q3	Give an advantage that warp-knitted fabrics have over weft-knitted fabrics.	[1 mark]
Q4	Suggest why chipboard shouldn't be used to make bathroom furnishings.	[1 mark]

Electronic Systems

A <u>system</u> is a collection of parts that <u>work together</u> to do a particular <u>function</u>. <u>Electronic systems</u> are made up of <u>components</u> that are connected to form a <u>circuit</u> — all electronic products have one...

All Systems have an Input, a Process and an Output

1) A system can be broken down into three 'blocks' — <u>input</u>, <u>process</u> and <u>output</u>.
2) A <u>signal</u> (e.g. electricity, movement, light) passes from one block to the next.
3) Each block <u>changes</u> the signal in some way.

It might be easier to understand these blocks using an example from a <u>mechanical system</u> (see p.28). A <u>bicycle</u> is a good example:

1) The <u>turning</u> of the cyclist's <u>legs</u> is the <u>input signal</u> to the <u>pedals</u> (the <u>input block</u>).
2) This is then processed in the <u>chain and sprockets</u> (the <u>process block</u>).
3) The <u>output block</u> is the <u>rear wheel</u>, which <u>propels</u> the bike <u>forward</u>.

Input → Process → Output

Sprockets are toothed wheels that are connected by a chain.

Electronic Systems Involve Circuits

1) <u>Circuits</u> are made up of <u>components</u> that are joined together by <u>wires</u>.
2) Lots of electronic systems, however, make use of <u>printed circuit boards</u> (PCBs). These are boards with <u>thin copper tracks</u> (instead of wires) connecting components in the circuit — they're used to <u>reduce</u> the <u>size</u> and <u>manufacturing cost</u> of electronic systems.
3) The materials you use in a circuit have to be <u>conductors</u> — they need to let electricity <u>flow through them</u>. <u>Copper</u> is used for the wire or tracks because it's <u>ductile</u>, <u>malleable</u> and a <u>good conductor</u>.
4) Circuits can be drawn using a <u>circuit diagram</u> — <u>symbols</u> are used to represent <u>components</u>, and the <u>straight lines</u> linking them represent how they're connected up (e.g. with wires).
5) <u>Voltage</u> from a power cell (e.g. a battery) or the mains <u>pushes</u> the electric <u>current</u> around a circuit.
6) <u>Resistors</u> can be used to <u>reduce</u> the current in a circuit so you don't damage delicate components (such as a light bulb). Resistance is <u>measured</u> in <u>ohms</u> (Ω). A <u>larger</u> resistance means <u>less current</u> in the circuit.
7) In electronic systems, the input, process and output blocks are the <u>circuit components</u>, e.g. a <u>switch</u>, <u>buzzer</u>, <u>lamp</u> etc.

Copper tracks

Component	Symbol
Battery	
Resistor	
Switch	open closed
Buzzer	
Bulb	
Speaker	

Colour-coded <u>stripes</u> show the <u>resistance</u>.

For example — an <u>electronic egg timer</u>:

Input signal → Input block SWITCH → Signal → Process block TIMER → Signal → Output block BUZZER → Output signal

You <u>press</u> the <u>switch</u>... ...which <u>completes</u> the circuit and <u>starts</u> the <u>timer</u>... ...which adds a set <u>time delay</u> to the system (see p.27). When this is up, it passes on the signal to the <u>buzzer</u>... ...which makes loads of <u>noise</u>.

Electronic Systems

Input Devices Change the Electrical Current in a Circuit

1) Input devices <u>receive</u> an <u>external</u> signal — this is what <u>triggers</u> a <u>system</u> to work.

2) They're often <u>sensors</u> — <u>components</u> that <u>convert</u> a specific type of <u>energy</u> into an <u>electrical signal</u>.

3) There are various types of <u>input device</u> that you can use in an electronic system. These include <u>switches</u> and <u>variable resistors</u>.

Switches Turn a Circuit On and Off

1) Turning a switch on <u>completes</u> a circuit. This <u>allows current to flow</u>.

2) There are many different types of switches — common ones include <u>toggle</u>, <u>push</u> and <u>slide</u> switches.

Toggle switch — flick to one position to complete the circuit.

Push switch — push down and hold to complete/ break the circuit.

Slide switch — slide to one side to complete the circuit.

Variable Resistors Can Detect Changes in External Conditions

1) Variable resistors <u>change the resistance</u> of the circuit, often based on an <u>external factor</u> such as temperature.

2) This alters the <u>current</u> in the wires, which gets passed on to the <u>process block</u>.

3) Here are some <u>common types</u> of variable resistor:

Thermistors

1) <u>Thermistors</u> detect changes in <u>temperature</u> — in <u>hot</u> conditions, the <u>resistance falls</u> and <u>lots</u> of <u>current</u> can flow through them.

2) They're used in <u>central heating systems</u> that <u>turn off</u> when the room reaches a <u>certain temperature</u>.

Light-dependent resistors

1) <u>Light-dependent resistors</u> (LDRs) are <u>light sensors</u> — they detect changes in <u>light levels</u>.

2) In <u>brighter light</u>, the <u>resistance</u> of an LDR <u>falls</u> — this allows circuits to be turned <u>on or off</u> when it gets <u>light or dark</u>.

3) For example, LDRs are often used in <u>automatic night lights</u> — the light is turned <u>on</u> when it gets <u>dark</u>.

Pressure Sensors

1) <u>Pressure sensors</u> detect changes in <u>pressure</u> in a <u>system</u>.

2) Depending on the <u>type</u> of pressure sensor, resistance can either <u>increase or decrease</u> as pressure increases.

In a sealed container or system, the more gas or liquid there is, the higher the pressure.

3) They're often used in equipment where <u>gases or liquids</u> are <u>monitored</u>. In <u>cars</u>, pressure sensors are used to detect if there's a <u>leak</u> in the <u>fuel system</u> — a leak would cause the pressure to <u>fall</u>, and this would <u>trigger</u> a <u>warning light</u> to come on.

I'm like an LDR — catch me in the right light and I'm irresistible...

Input devices are exactly what they say on the tin — they allow an input into a system. The system then has to make a decision based on the input (or inputs), but we'll get on to this on the next page...

Q1 Name the three blocks that make up a system. [3 marks]

Q2 Describe the difference between a signal and a block in a system. [2 marks]

Q3 Describe what a printed circuit board (PCB) is. [2 marks]

Q4 a) What is the function of a variable resistor in a circuit? [2 marks]

b) Give three examples of types of variable resistor. For each resistor, give the external factor that it can detect changes in. [3 marks]

c) Give an example of where a named type of variable resistor can be used. [1 mark]

Electronic Systems

Now it's time for the juicy part of electronic systems — processing the inputs and producing outputs.

Process Devices Receive Inputs and Decide the Outputs

Process devices make all the decisions in an electronic system — they process the input information and use it to determine the output. Devices called integrated circuits can carry out processing in electronic systems.

Integrated Circuits Combine Many Electronic Components

1) Integrated circuits (ICs) or 'chips' are tiny, self-contained circuits. They can have billions of components built in, so they can get very complicated.

2) ICs help to simplify electronic systems by reducing the number of separate components that are needed in a circuit. This means that systems with ICs are:

  • Cheaper to make (as lots of separate components can be expensive).

  • Smaller — this is particularly important in portable electronic products, e.g. mobile phones.

  • Use much less power (because of their size).

There are loads of different types of integrated circuit — microcontrollers are just one type of IC.

3) You can treat ICs as a single process block.

Microcontrollers are a Special Type of Integrated Circuit

1) A microcontroller is a mini computer on a chip — it has a processor, memory, and one or more inputs and outputs.

2) They're found in pretty much all electronic products, e.g. microwave ovens, washing machines etc.

Advantages
1) Microcontrollers can do the jobs of multiple ICs — this simplifies the system even further.
2) Unlike other ICs, most can be reprogrammed for a different use.

Disadvantage
Microcontrollers are more expensive than other ICs.

A commonly used microcontroller is a PIC, a programmable intelligent computer (see p.89).

Microcontrollers are Programmed for a Certain Task

1) Microcontrollers need to be programmed (given instructions) to perform a particular function — without this, they won't do anything at all. The program is stored in the microcontroller's onboard memory.

2) Often the programming of microcontrollers is carried out when they are manufactured and the program isn't altered after this. However, it is possible to program microcontrollers yourself.

3) There are several microcontroller boards (a PCB with a microcontroller along with other components) that can be programmed for use in a variety of projects — e.g. Arduino boards and the BBC micro:bit.

4) Programming can involve either:

  • Writing a series of commands using a programming language — examples include BASIC and Scratch.

  • Creating a flowchart telling the microcontroller exactly when to start, how to make decisions based on input signals, when to add a time delay to the process, what to output, and when to stop operating.

Different shaped boxes represent different stages in a flowchart.

5) This is done using specialist computer software such as PICAXE (which uses BASIC) and Logicator (which uses flowcharts) or online tools such as the BBC micro:bit website. The instructions are transferred to a microcontroller so they can be used.

6) Microcontrollers can be programmed for a range of different tasks. You need to know about them being programmed as timers, counters and for decision-making — see next page.

Electronic Systems

Microcontrollers as Timers and Counters

1) Timers and counters have <u>similar functions</u> — they can both be involved in <u>timing</u> in circuits. They both work by responding to <u>pulses</u> (peaks of higher <u>voltage</u>) in circuits.

2) <u>Timers</u> are often used to <u>add</u> a <u>time delay</u> to a process and make sure parts of a program are running when they're supposed to — they can <u>generate a pulse</u> after a particular length of time has passed.

3) <u>Counters</u> differ from timers because they <u>count pulses</u> of voltage produced by an <u>input device</u>.

4) Microcontrollers usually contain timers and counters <u>inside them</u>, which allow the programs they run to include <u>timing</u> and/or <u>counting</u> functions.

Microcontrollers as Timers

- Microcontrollers controlling <u>flashing lights</u>, like car indicators, use timers to set the time that the light is on and off for.
- Microcontrollers in <u>microwaves</u> use a timer to set the length of time that the oven cooks for.

Microcontrollers as Counters

Most modern cars have a dial on the dashboard that shows <u>how fast</u> the engine is <u>turning</u> — the number of <u>revolutions per minute</u>. A microcontroller is used to <u>count</u> the number of revolutions.

Microcontrollers as Decision-Makers

1) <u>Logic gates</u> make <u>decisions</u> in <u>ICs</u> based on a collection of <u>inputs</u>.
2) <u>Inputs</u> must be <u>digital</u> — they're either <u>on</u> (1) or <u>off</u> (0).
3) The names of logic gates are, well, pretty logical...

- <u>OR gates</u> have <u>2 inputs</u>: if one input <u>or</u> the other is <u>on</u> (or <u>both</u>), then the output is <u>on</u>. An <u>automatic door</u> is a good example — the door will open if a <u>sensor detects</u> someone on the <u>outside or inside</u>.
- <u>AND gates</u> have <u>2 inputs</u>: <u>both</u> inputs need to be <u>on</u> for the output to be <u>on</u>. <u>Tumble dryers</u> won't start until <u>both</u> the <u>start button</u> is pressed <u>and</u> an <u>input sensor</u> has detected that the <u>door</u> is <u>closed</u>.
- <u>NOT gates</u> have <u>1 input</u>: if the input is <u>on</u>, the output is <u>not on</u>. It <u>reverses the input</u>. Pressing an <u>emergency stop switch</u> will stop the output — on an <u>escalator</u>, this would cause it to <u>stop moving</u>.

Output Devices Include Lamps, Buzzers and Speakers

1) The <u>output device</u> you use in a system depends on what your product has to <u>do</u>. It determines how the system <u>responds</u> to the input signal.

2) There are absolutely <u>loads</u> of output devices... here are a few you <u>need to know</u> for the exam:

<u>LEDs</u> (light-emitting diodes) and <u>lamps</u> (e.g. bulbs) turn electricity into <u>light</u>, e.g. <u>security lights</u>.

<u>Buzzers</u> make a <u>noise</u>. They're used in <u>alarm clocks</u> to wake you up at stupid o'clock...

<u>Speakers</u> turn electrical signals into <u>sound</u>, e.g. in <u>headphones</u>.

All this talk of inputs and outputs — it's like the hokey cokey...

"Shake it all about" isn't an acceptable example of a process though... unfortunately. Remembering some simple systems can really jog your memory when it comes to understanding what makes an input, process and output.

Q1 What is an integrated circuit? [1 mark]
Q2 Give two advantages of using microcontrollers in a system. [2 marks]
Q3 Give two ways of programming a microcontroller. [2 marks]
Q4 Give two uses of timers. [2 marks]
Q5 Describe how an AND logic gate works. [1 mark]
Q6 Name two output devices and state the type of output signal that they produce. [2 marks]

Mechanical Systems

Electrical systems aren't the only type of system you need to know about. There are also mechanical systems.

Mechanisms Change the Magnitude and Direction of Force

1) All mechanical systems have mechanisms which transform an input motion and force into an output one.

2) They're designed so you can gain a mechanical advantage from using them — they make something easier to do. This often involves changing the magnitude (size) and direction of the applied force.

3) Some mechanisms change one type of motion into another.

4) There are various types of motion:

- Linear motion — moving one way in a straight line.
- Reciprocating motion — moving backwards and forwards in a straight line, e.g. a piston.
- Oscillating motion — moving backwards and forwards in an arc, e.g. a swing.
- Rotary motion — moving in a circle, e.g. a wheel.

For example, a car jack lets you lift up a car — a job you couldn't do without it. You turn a handle (rotary motion) with a relatively small force and the car is lifted upwards (linear motion) by a large force.

Compressed air goes in here... ...and forces the piston out. The spring moves the piston back.

Levers Make It Easier to Move and Lift Things

Levers move and lift loads by rotating about stationary points called fixed pivots. There are three main types of lever that you need to know about.

First Order Levers Have a Pivot in the Middle

1) First order levers have the pivot between the effort and the load. Pushing down on your side of the lever lifts the load up at the other end — the lever has rotated about the pivot.

2) If the load is closer to the pivot than the effort, a large load can be lifted using a smaller effort — the lever gives a mechanical advantage.

3) As you move the pivot closer to the load it becomes easier to lift — the magnitude of the effort required decreases.

First order levers can make it much easier to lift heavy loads — you could even lift a sumo or two if they were close enough to the pivot. Try it if you don't believe me...

Second Order Levers Have the Load in the Middle

1) Here the pivot is at one end of the lever and the effort is at the other end. The classic example is a wheelbarrow.

2) The closer together the pivot and load are, the easier it is to lift.

Third Order Levers Have the Effort in the Middle

1) In a third order lever the effort is in between the load and the pivot.

2) Third order levers can be things like fishing rods, cricket bats and garden spades.

3) Moving the effort and pivot further apart makes it easier to move or lift the load.

If only there was a mechanism to make revision easier...

Mechanisms are all about making things easier to do. Levers can change the magnitude of forces — a small input force (effort) can lift a large output (load) by using a simple pivot. Pretty useful stuff I'm sure you'll agree...

Mechanical Systems

Linkages Connect Different Parts of a Mechanism

Levers can be connected together to form linkages. Simple linkages can change the magnitude of the force and the direction of motion. Here are two examples:

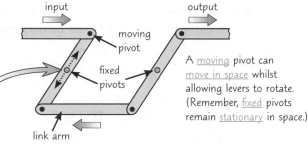

Push/Pull Linkages
1) Push/pull linkages use two fixed pivots.
2) The input and output motions of the linkage are in the same direction. The motion of the link arm is in the opposite direction.
3) Here, each fixed pivot is in the centre of an arm. Changing the position of these pivots will change the magnitude of the output force — often the easiest way to see this is by making a model.

A moving pivot can move in space whilst allowing levers to rotate. (Remember, fixed pivots remain stationary in space.)

Bell Crank
1) A bell crank changes the direction of a force through 90°.
2) The magnitude of the output force can be changed by moving the fixed pivot so it's not an equal distance between the two moving pivots.

Gear Trains Transmit or Change Rotary Motion

1) Gears are toothed wheels that interlock. They transfer motion from one part of a machine to another.
2) A gear train is where two or more gears are linked together. They can be used to change the direction of motion or change the magnitude of the input force (this is to do with the increase or decrease in the speed of rotation that can be generated by linking different-sized gears). Here are some examples:

The driver gear, turned by hand or a motor, turns the driven gear.
Both will turn in opposite directions.

Driver • • Driven

If you use a third gear (called an idler), the driver and the driven gears will both turn in the same direction. The size of the idler won't alter the speed of the other two gears.

Idler

If linked gears are different sizes, the smaller gear (i.e. the one with fewer teeth) will turn faster. The relationship between the driver and the driven gears can be described using a gear ratio. The size of this ratio describes how much the mechanism changes the speed of the gears from the input speed to the output speed. Here's an example of how gear ratios can be used in calculations:

$$\text{Gear ratio} = \frac{\text{no. of teeth on the driven gear}}{\text{no. of teeth on the driver gear}}$$

$$\text{Output speed} = \frac{\text{speed of driver gear (input speed)}}{\text{gear ratio}}$$

EXAMPLE:

A gear train is made up of a driver gear with 10 teeth and a driven gear with 20 teeth. The driver gear is rotating at 500 rpm (revolutions per minute). Calculate the output speed of the gear train.

Driver 10 teeth   Driven 20 teeth

Gear ratio = 20 ÷ 10 = 2/1 or 2:1 or 2   Output speed = 500 ÷ 2 = __250 rpm__

Gears — now there's a topic to get your teeth into...

Q1 a) Describe reciprocating motion and give an example of where it can be found. [2 marks]

  b) Describe oscillating motion and give an example of where it can be found. [2 marks]

Q2 State an example of a second order lever. [1 mark]

Q3 A gear train contains two gears. The driver gear has 10 teeth and the driven gear has 50 teeth. Calculate the gear ratio of this gear train. [2 marks]

Mechanical Systems

Rotary systems are mechanical systems that work by using rotary motion. Gear trains on the previous page are an example of a rotary system but there are a few more that you need to know about...

Pulleys Help to Lift a Load

1) A simple pulley is made up of a wheel with a grooved outer edge and a cable, rope or belt (see below) that sits in this groove.

2) Pulleys make lifting a load easier.

Cranes use pulley mechanisms. Or wings.

One pulley doesn't give any mechanical advantage.

3) One pulley on its own changes the direction of the force required. The same amount of force is needed but pulling down might be easier than lifting something up.

Simple block and tackle

Weight can be lifted using half the effort.

4) Using two or more pulleys together can change the magnitude of the force too — they can make things feel a lot lighter than they actually are (if you set them up correctly). For example, one fixed pulley and one moving pulley (a block and tackle) will mean you only need half the force to lift a load.

Belt Drives Transfer Movement

1) A belt drive transfers movement from one rotating shaft to another.

2) Belt drives are used in pillar drills. They're also used in products such as washing machines (see p.86).

3) In a pillar drill, a flexible belt joins two separate pulley wheels — this links the motor to the drill shaft.

Belt

motor (driver)

drill shaft (driven)

pillar drill

4) The belt can be put in different positions to make the drill turn faster or slower. This works in a similar way to gears (see previous page) — if the wheels are of different sizes, the smaller wheel will spin faster.

5) The diameter (width) of the wheels can be used to calculate the velocity ratio — how fast the driven wheel will spin relative to the driver wheel:

A similar mechanical system is a chain and sprocket. These are made of toothed wheels (called sprockets) connected together by a chain. They're used on bikes.

$$\text{Velocity ratio} = \frac{\text{diameter of the driven pulley wheel}}{\text{diameter of the driver pulley wheel}}$$

6) You can use another formula to calculate the output speed of the driven pulley wheel.

$$\text{Output speed} = \frac{\text{speed of driver pulley wheel (input speed)}}{\text{velocity ratio}}$$

This formula is very similar to the one we saw for gears on the previous page.

Rpm means revolutions per minute — it's a measure of the speed of rotation.

EXAMPLE:

A belt drive is made up of a driver wheel with an 30 mm diameter and a driven wheel with a 90 mm diameter. The driver wheel is rotating at 600 rpm. Calculate the output speed of the driven wheel.

Velocity ratio = 90 ÷ 30 = 3/1 or 3:1 or 3   Output speed = 600 ÷ 3 = 200 rpm

---

A block and tackle — seems like P.E. all of a sudden...

Yep, there are all sorts of useful mechanical systems on these pages. But fear not — take them one at a time and make sure you understand what's going on in the diagrams. You'll have it sorted before you can say belt drive...

Mechanical Systems

Cams Change Rotary Motion to Reciprocating Motion

1) A <u>cam mechanism</u> has <u>two main parts</u> — the <u>cam</u> and the <u>follower</u>.

2) Cams come in many different shapes and sizes. They always <u>rotate</u>.

3) The <u>follower</u> rests on the cam as it rotates, and <u>follows its shape</u>. It may have a small wheel to reduce <u>friction</u>. The follower moves <u>up and down</u> (reciprocating motion) as the cam turns.

WHEEL FOLLOWER

CAM

rotates

Here are a few basic <u>cam shapes</u>:

① CIRCULAR CAM (also called <u>offset</u> or <u>eccentric</u>) — produces a <u>uniform reciprocating motion</u>.

The circular cam rotates about an off-centre pivot...

... which causes the follower to move up and down.

② SNAIL CAM — For half a turn the follower will not move, then it will <u>gently rise, and then suddenly drop</u>. It will only work in <u>one direction</u>.

Rotates in one direction only

③ PEAR CAM — Again for half a turn the follower will not move, then it will <u>gently rise and fall</u>.

④ FOUR-LOBED CAM — Has four lobes (bits that stick out). For <u>each turn</u> of the cam the follower will <u>rise and suddenly fall four times</u>. This cam shape will also only work in <u>one direction</u>.

Rotates in one direction only

4) Changing the <u>size</u> or <u>shape</u> of the <u>cam</u> can be used to <u>change</u> the <u>magnitude</u> of the <u>output</u> motion (the <u>reciprocating</u> motion of the <u>follower</u>).

Lots to learn — but you cam do it...

Q1 a) What is the name of the mechanical system shown on the right? [1 mark]
    b) What is the purpose of this mechanical system? [1 mark]

Q2 A pillar drill operates using a belt drive mechanism. The driver wheel has a diameter of 32 mm and spins at a speed of 1200 rpm. The driven wheel has a diameter of 128 mm.
    a) Calculate the velocity ratio. [2 marks]
    b) Calculate the output speed of the system. [2 marks]
    c) Name one other product that uses a belt drive mechanism. [1 mark]

Q3 Describe what a simple pulley mechanism is made up of. [2 marks]

Q4 Nadia needs a mechanism that will help her to lift a heavy load of bricks.
    a) Suggest a mechanical system that could help with this. [1 mark]
    b) Explain how it helps. [1 mark]

Q5 Phillip is designing a product which includes a cam.
    a) Describe the motion produced by the cam shown on the right. [1 mark]
    b) Why do some followers have a small wheel? [1 mark]
    c) Name the type of cam shown on the right. [1 mark]

Developments in New Materials

New materials are being developed all the time. They can have some seriously snazzy properties too...

Modern Materials are... well, Modern

1) Modern materials are normally materials that have been developed with a specific application in mind.

2) They can be developed by inventing new or improved manufacturing processes. For example:

Graphene

1) Graphene is a super-thin layer of graphite — the stuff used in pencil leads.

2) It has some amazingly useful properties — it's incredibly light and strong and is a great conductor of heat and electricity.

3) It was first made in 2004. Since then, several manufacturing processes have been invented and improved to allow it to be made more cheaply.

4) Graphene can be used in modern tennis rackets. Future applications include aerospace, vehicles, water purification and flexible electronics.

Metal Foams

1) Metal foam is a metal (e.g. aluminium) that contains many gas-filled spaces which make the material lightweight. These foams also keep some of the metal's properties too — they're stiff, tough, and strong under compression.

2) Metal foam was developed in the 1940s. Research since has focussed on improving its properties and production methods.

3) These foams are already being used for lightweight car parts and in bone implants.

Titanium

1) Titanium is an extremely corrosion-resistant metal with a high strength-to-weight ratio.

2) A process was developed in the 1930s that allowed titanium to be extracted easily and cheaply enough to be used in the aerospace industry (aircraft and spacecraft) — this is its main use to this day.

3) Titanium is difficult and therefore expensive to machine (e.g. drill, cut, polish). Recent developments in CAD/CAM have helped to reduce machining costs and make it a more usable material, e.g. for bone replacements, dental implants, bikes, ships and armour.

3) Some modern materials can be developed by altering an existing material to perform a particular function:

Liquid Crystal Displays (LCDs)

1) LCDs are used in flat-screen displays that are thin, lightweight and energy efficient.

2) The liquid crystals used in the display are made of a mixture of chemicals. When an electric current is applied, the crystal's shape is modified — this in turn changes the image seen on the screen.

3) LCDs were originally used in calculators and other small displays and were usually just black and grey. They have developed to become high-definition, full-colour displays used as TV and computer screens.

Coated Metals

1) By coating metals with a different material, their properties can be altered.

2) E.g. iron and steel can be galvanised — coated with zinc to prevent rusting (see p.73). Electroplating can also be used to achieve this, e.g. nickel-plated steel is a cheaper, corrosion-resistant alternative to stainless steel in car parts.

3) Anodised aluminium has a coating of aluminium oxide to make the surface harder and more resistant to corrosion (see p.87).

4) Metals (e.g. steel) can be coated with PVC (see p.19) — used for roofing, the PVC makes it corrosion-resistant and can be coloured.

Nanomaterials

Nanomaterials are made of tiny particles (nanoparticles). These particles have always existed but our ability to manipulate them for specific purposes is a fairly recent development. Nanomaterial uses include:

1) Carbon nanotubes are tiny carbon cylinders. They have a very high strength-to-weight ratio, and are good conductors of heat and electricity. They can be added to a material to strengthen it without adding much weight (e.g. in tennis racquets). They're also used in electronics.

2) Self-cleaning fabrics have a nanoparticle coating that removes odours and stains upon exposure to light.

3) Antibacterial fabrics use nanoparticles of silver to kill bacteria. They have lots of medical uses, e.g. face masks and dressings. They're also used in anti-bacterial toys and odour-free socks.

Developments in New Materials

Smart Materials React to Their Environment

Smart materials <u>change</u> their <u>properties</u> in response to stimuli, e.g. <u>temperature</u>, <u>light</u>, <u>stress</u>, <u>moisture</u> or <u>pH</u>. They often <u>return</u> to their <u>original state</u> when the stimulus is <u>taken away</u> — the changes are <u>reversible</u>.

1) <u>Shape memory alloys</u> are alloys that 'remember' their <u>original shape</u>. They can be easily shaped when cool, but they <u>return</u> to their <u>original shape</u> when <u>heated</u> above a <u>certain temperature</u>.

2) An example of a shape memory alloy is <u>nitinol</u>. If your <u>glasses</u> are made of nitinol and you accidentally <u>bend</u> them (stress), you can just pop them into a bowl of hot water and they'll <u>jump</u> back <u>into shape</u>.

---

1) <u>Photochromic pigments</u> change colour <u>reversibly</u> in response to <u>light</u>.

2) They can be put into <u>spectacle lenses</u> to make glasses that turn into <u>sunglasses</u> when it's sunny.

3) <u>Photochromic inks</u> can be used to print t-shirts with <u>designs</u> that only <u>show up in sunlight</u>.

---

1) <u>Thermochromic pigments</u> and inks are used in <u>colour changing</u> products — they react to temperature.

2) When the <u>temperature</u> changes, the product <u>changes colour</u>. The colour <u>changes back</u> when the object returns to its <u>original temperature</u>.

3) They're used in <u>babies' feeding spoons</u> so the parent knows the food isn't too hot, and in <u>novelty mugs</u> and <u>t-shirts</u>.

> Thermochromic ink can be used to print thermochromic plastic sheets, which change colour as the temperature changes. This is used in some thermometers.

Blue spoon   gets hot...   ...and becomes pink

Composites are a Combination of Two or More Materials

<u>Composites</u> are made from <u>two or more different materials</u> bonded together. These enhanced materials often have <u>different</u> (more <u>useful</u>) <u>properties</u> than those of the <u>individual materials</u> they're made from. For example:

Composite	Made from	Properties	Uses
Glass reinforced plastic (GRP)	Glass fibres that are coated in a thermosetting plastic resin.	Stronger and tougher than the plastic by itself, heat-resistant, easy to mould into complex shapes.	Boats, kayaks, surfboards, some car bodies, PCBs
Carbonfibre reinforced plastic (CRP)	Carbon fibres that are coated in a thermosetting plastic resin.	Lighter, tougher and stronger than GRP, but more expensive.	Protective helmets, racing cars, sports equipment, laptops, bulletproof vests

Technical Textiles are Made Just to be Functional

<u>Technical textiles</u> are <u>enhanced fabrics</u>. They're <u>designed purely</u> to be <u>functional</u>, rather than <u>look good</u>. E.g:

1) <u>Kevlar®</u> — a <u>synthetic fibre</u> that can be woven into a <u>really strong</u> fabric that's <u>resistant to abrasion</u>. This makes it useful, e.g. in <u>bulletproof</u> vests, clothing for <u>motorcyclists</u>, to reinforce <u>tyres</u>, etc.

2) Some <u>synthetic fibres</u> are designed to be very <u>fire-resistant</u>, e.g. <u>Nomex®</u>. These enhanced fabrics are <u>different</u> to those <u>treated</u> for flame retardance (see p.74), as Nomex® has fire-resistance <u>built into the fibres</u>, so it <u>can't</u> be <u>washed</u> or <u>worn away</u>. They're used in <u>firefighters'</u> and <u>racing drivers'</u> overalls.

3) <u>Micro encapsulation</u> — this is where <u>tiny droplets</u> of a <u>chemical</u> are coated in <u>shells</u> called <u>microcapsules</u>, and embedded in a <u>microfibre fabric</u> (one woven from very thin synthetic fibres). These chemicals have a <u>functional</u> use, e.g. as <u>insect repellent</u>, an <u>odour neutraliser</u>, or a <u>perfume</u>. They're used in products such as <u>antibacterial socks</u> and <u>scented lingerie</u>.

4) <u>Conductive fabrics</u> — use <u>fibres</u> that can <u>conduct electricity</u>. They're used to <u>integrate electronics into clothing</u> and in <u>touchscreen gloves</u>.

Smart and modern... Well I don't like to brag...

Remember, modern materials have useful properties built in, smart materials react to changes in the environment.

Q1	Name two properties of graphene.	[2 marks]
Q2	What material are the nanoparticles made of in antibacterial fabrics?	[1 mark]
Q3	State two uses of glass-reinforced plastic.	[2 marks]

Selecting Materials

When <u>choosing</u> a <u>material</u> to use in a product, there's a lot to consider. There's <u>rarely</u> a <u>perfect</u> material — so you'll have to <u>weigh up</u> which <u>factors</u> are the <u>most important</u> and <u>prioritise</u> those first...

There are Many Factors to Think About...

Functionality

1) Probably the most <u>important factor</u> is whether the material is <u>functional</u> — it must have <u>properties</u> that <u>help</u> it to <u>do the job</u> it's <u>supposed to</u> in a product.

2) You need to have an idea of <u>how</u> the product <u>will be used</u> and the <u>demands</u> that will be made on the material. For example, if you were making a bed for <u>elephants</u>, you'd need materials <u>strong enough</u> to support their weight (e.g. steel).

3) You also need to consider how <u>easy</u> the <u>materials</u> are <u>to work with</u>. E.g. you wouldn't want to make a detailed wooden carving out of <u>pine</u> (see p.17) — it's <u>knotty</u>, so would be <u>hard to shape</u>.

Availability

1) You need to consider <u>how easy</u> it is to <u>source</u> (find) and <u>buy</u> the materials you want in a <u>suitable form</u>.

2) <u>Widely available</u> materials are <u>ideal</u> — you can get your hands on them <u>quickly</u> and <u>easily</u>. <u>Less widely available</u> materials are often more <u>expensive</u>. They also may need to be <u>delivered</u> from <u>further away</u>, which adds to the <u>cost</u> and <u>waiting time</u>.

3) A lot of materials are <u>only widely available</u> in <u>stock forms</u> and <u>sizes</u> (see p.60). You should try to use these <u>stock</u> forms and sizes if possible, as it can be very <u>expensive</u> to get them in <u>other forms</u>.

Aesthetics

1) If a material is going to be <u>seen</u>, you want it to <u>look good</u>. You should <u>consider</u>:
   - <u>Colour</u> — e.g. using a thread that's the same colour as a fabric may look better on a garment.
   - <u>Surface finishes</u> — e.g. <u>painting</u>, <u>varnishing</u>, <u>polishing</u>, <u>lacquering</u> etc. all <u>change</u> the <u>appearance</u> of a material, as well as helping to <u>protect</u> it from moisture and dirt.

2) You also need to think about a material's <u>texture</u> — how a material <u>feels</u> to the touch, e.g. rough, soft etc. This is especially important in <u>textiles</u>, when the product will be <u>worn</u>.

3) The aesthetics must <u>appeal</u> to your <u>target market</u>. For example, <u>traditional</u> oak furniture might appeal to <u>older people</u>, whereas younger people may prefer <u>brightly coloured</u> polypropylene.

Cost of Materials

When working out <u>how much</u> you can <u>spend</u> on <u>raw materials</u>, you should always consider:

1) <u>How much</u> you're going to <u>sell</u> the <u>product</u> for. You <u>shouldn't spend loads</u> on expensive <u>materials</u> if you're <u>selling</u> the product <u>cheaply</u> — you'll just <u>lose money</u>.

2) The <u>amount of each material</u> you're going to <u>use</u>. <u>Affordable products</u> can still use fairly <u>expensive materials</u> in <u>small quantities</u>. For instance, a <u>gold-plated</u> photo frame uses far less gold than a <u>solid gold</u> photo frame, so can be <u>made</u> and <u>sold</u> at a much <u>cheaper</u> price.

Ruby encrusted solid gold photo frame — only 50p!

3) <u>How many products</u> you're making. <u>Mass-produced</u> products are <u>cheaper to make</u> than <u>one-off</u> products, because the <u>manufacturing cost per item</u> gets <u>lower</u> the <u>more you produce</u>. One of the reasons for this is that raw materials can be <u>bought in bulk</u> — because you're buying so much, it allows you to <u>negotiate a discount</u> with the <u>supplier</u>.

Aesthetics, costs — this designing lark is so materialistic...

You'll need to find a compromise between factors, e.g. using cheap materials might affect aesthetics, so you may need to cut costs in a different way. These aren't the only factors to consider though — there's more coming up...

Selecting Materials

Environmental Factors

1) Where possible, you should try to use sustainable materials to limit the environmental impact of the product, e.g. materials that are biodegradable or that come from renewable resources (see p.6-8).

2) However, a lot of the materials used in products aren't very sustainable, for example...

Metal	• Comes from non-renewable metal ores • Uses a lot of non-renewable energy to mine it • Mining damages ecosystems
Plastic	• Non-renewable (made from oil) • Not biodegradable
Paper/ Wood	• Renewable, but only if trees are replanted to replace the ones cut down • Deforestation (see p.47) damages ecosystems

3) It's not all bad though — some of these materials can be:
   • Recycled (for a different use), e.g. thermoforming plastics can be melted and remoulded (see p.19).
   • Re-used (for the same use), e.g. plastic bags can be re-used many times before breaking.

4) Choosing to use recycled or re-used materials in a product helps to limit the amount of non-biodegradable stuff on rubbish tips and the amount of non-renewable material that needs to be obtained.

There's more about the processes involved in sourcing these materials on p.44-47.

Social Factors

1) As a designer, you have a social responsibility — a duty to act in a way that benefits society and the environment. This means you shouldn't select materials based only on which ones will make the most money. Instead, you may need to compromise on factors such as the cost or availability of materials.

2) For example, using Fairtrade materials (see p.103) may cost more than using materials from a less ethical source, but it may be considered worthwhile to have a positive social impact on the lives of the farmers and workers who provide the materials.

3) Using recycled materials in products can cost more, but there are social benefits:
   • To obtain raw materials, non-renewable energy sources are often used, e.g. the burning of fossil fuels. This causes air pollution, which can result in long-term health problems such as lung disease and cancer. Although recycling causes some air pollution, it usually doesn't cause as much as the processes used to obtain raw materials. Recycling can therefore benefit the health of society.
   • Obtaining new materials also requires large amounts of land use, e.g. to mine coal and metal, to drill for onshore oil and gas, and to chop down trees. This can mean a reduction in the quality of life for those who live nearby, relocation of communities and the destruction of livelihoods. Recycling helps to reduce the number of people that are impacted by the materials industry.

Ethical Factors

1) Ethically sourced materials are produced in an environmentally sustainable way that is also fair to workers.

2) Some materials have logos to show they have met certain ethical standards. E.g. the logo of the Forest Stewardship Council® (FSC®) shows that the timber (or paper) in a product has come from responsibly managed forests and/or verified recycled sources.

3) Animal products such as fur, ivory and leather are often viewed as unethical, as animals are killed for the materials. Synthetic materials could be considered as an ethical alternative, e.g. artificial fur.

Cultural Factors

Culture covers everything from religion, beliefs and laws to languages, food, dress, art and traditions.

1) If you're designing a product aimed at a specific target market, you'll have to take into account the views and feelings of people from that particular culture.

2) For example, using transparent fabrics in garments may offend people in some cultures, and in China, black can be seen as bad luck.

Recycling — what a load of rubbish...

Q1   Give three factors that you should consider when selecting a material to use in a product.   [3 marks]

Q2   Suggest two disadvantages of using a material that is not widely available.   [2 marks]

Q3   Isaac is hand-crafting a one-off chair made from expensive mahogany.
      Suggest two reasons why Isaac needs to sell it at a high price.   [2 marks]

Forces and Stresses

Forces and stresses are part of everyday life — stuff is always being stretched, squashed, bent and twisted. Objects are designed to cope with these forces, and in some cases, they can use them to their advantage...

Forces Can Act on Objects in Different Ways

1) Forces can cause objects and materials to break or change shape. Materials are strong if they are good at withstanding a force without breaking (see p.14). Force is measured in newtons (N).

2) Stress is the force per unit area. It's often measured in N/m^2.

3) There are a few different types of forces and stresses that you need to know about:

Tension

1) Tensile forces act to stretch an object and pull it apart: ←[ ]→
E.g. in a tug-of-war, the rope is under tension caused by people pulling in opposite directions.

2) Trampolines use springs with a high tensile strength — when you land on a trampoline, you stretch the springs. The springs resist being permanently stretched and try to return to their original length — this throws you back up into the air.

3) Suspension bridges use cables with high tensile strength to support the weight of the bridge.

If these things didn't have a high tensile strength, they'd snap as a result of the forces acting on them.

Cables under tension

Bridge support under compression

Compression

1) Compression is the opposite of tension. Compressive forces are directed towards one another, and act to squash and shorten objects: →[ ]←

2) Compression often occurs when a material is supporting a weight above it. Chair legs support the person sitting on the chair and bridge supports hold the weight of the bridge. They're made from materials that have a very high compressive strength.

Shear

1) Shear forces often act in opposite directions but unlike compressive and tensile forces, they aren't aligned with one another. This can cause parts of an object to slide past one another.

2) For example, scissors use shear forces to cut (shear) materials.

3) Industrial cutting machines such as guillotines (see p.52) and die cutters (see p.53) also use shear forces to cut materials.

Forces on the top and bottom surfaces act in opposite directions.

The scissor blades push the paper in opposite directions, causing it to be ripped or cut.

Bending

Load

Bench legs push up on the bench.

1) A bending force is a type of shear force that causes materials to bend.

2) This commonly occurs when a load is applied to a certain part of a material — e.g. a bench has to resist bending forces when someone sits down in a certain place.

Torsion

1) Torsion acts to twist objects and materials. The forces attempt to rotate the different parts of a material in opposite directions.

2) Propeller shafts connect a ship's engine to the propeller, which spins in the water and drives the ship forwards. These shafts have to be able to resist torsion — it allows them to transfer rotary motion without breaking or twisting.

Engine rotates the propeller shaft

Propeller shaft

Water resists the rotation of the propeller and shaft.

The water applies a rotary force in the opposite direction to the engine — so a twisting force is applied to the shaft.

Forces and Stresses

Materials Can be Reinforced, Stiffened or Made More Flexible

Materials can be <u>reinforced</u> (strengthened) or <u>stiffened</u> (made more rigid) to help <u>resist</u> certain types of <u>forces</u>. They can also be <u>made more flexible</u> to better <u>work with forces</u>. <u>Enhancing</u> materials like this can <u>improve</u> their <u>functionality</u>. Here are a few ways in which materials can be <u>enhanced</u>...

> Rigid materials don't deform easily — they're the opposite of flexible materials. Rigidity is different to strength — strong materials don't break easily.

1) <u>Lamination</u> is a process in which one or more <u>layers</u> are <u>added</u> to a material to form a <u>composite</u> (see p.33).

2) It's often used to <u>increase strength</u> and <u>rigidity</u>.

3) For example, <u>foam core board</u> and <u>corrugated card</u> (see p.16) are <u>layered</u> — they're <u>stronger</u> and <u>stiffer</u> than other types of cardboard with a single layer of material.

4) <u>Plywood</u> is made up of <u>layers</u> of wood that are <u>glued</u> together to <u>resist bending forces</u>.

5) <u>Delicate fabrics</u> can also be laminated for extra <u>strength</u>, <u>stiffness</u> and <u>protection</u> (see p.75).

→ = Grain direction

The grain direction in a layer of plywood is at 90° to the layers above and below. This strengthens the weak lines of the grain.

<u>Interfacings</u> are <u>extra layers</u> of fabric <u>stuck</u> or <u>sewn</u> onto the <u>inside</u> of <u>products</u>. They're used in <u>collars</u>, <u>cuffs</u>, <u>pockets</u>, <u>waistbands</u> and <u>button holes</u> — anywhere that needs extra <u>strength</u> and <u>rigidity</u>. Interfacings are used to <u>improve functionality</u> and <u>aesthetics</u>.

<u>Webbing</u> is a <u>fabric</u> (usually cotton, nylon, polyester or polypropylene) that is <u>woven</u> in a way that gives a <u>very high tensile strength</u>. It's used in situations where it will be <u>under tension</u>, e.g. in <u>tow ropes</u>, <u>climbing harnesses</u>, <u>parachutes</u>, <u>seatbelts</u> and <u>backpack straps</u>.

<u>Bending</u> is often used to <u>reinforce</u> and <u>stiffen</u> materials. For example, the middle <u>fluted layer</u> of <u>corrugated card</u> is made from a <u>series of bends</u> that add <u>strength</u> and <u>rigidity</u> — important for packaging <u>heavy loads</u>.

<u>Folding</u> is the <u>bending</u> of a flexible material (e.g. <u>paper</u>, <u>card</u> or <u>fabric</u>) so that the <u>two sides</u> of the bend are <u>flat against each other</u> and a <u>line</u> called a <u>crease</u> or <u>fold</u> occurs <u>between them</u>. <u>Along a fold</u>, a material is <u>made more flexible</u>.

> Try folding a thick piece of cardboard... not that easy, right. Now try bending it along the fold you've just made — this should be loads easier.

The Laminator — part machine, part sheep...

You could be asked to describe how a material can be strengthened or reinforced, so make sure you've got some examples in mind. And if you get tension and torsion mixed up, remember that torsion involves **r**otation.

Q1   Which type of force acts to squash or shorten an object?   [1 mark]

Q2   The photo on the right shows someone abseiling down a wall. What type of force is acting on the climbing rope?   [1 mark]

Q3   Name one type of industrial cutting machine that uses shear to cut materials.   [1 mark]

Q4   Describe how you could apply torsion to a rod.   [1 mark]

Scales of Production

The <u>scale of production</u> is <u>how many</u> of a product you want to <u>make</u>. This decides the <u>type</u> of <u>production</u> that is <u>best suited</u> to the job. There are <u>four</u> main types of production that you need to know about...

One-Off Production Makes One Product at a Time

1) One-off production makes <u>bespoke</u> products — <u>every item</u> will be <u>different</u>, in order to meet a <u>customer's</u> exact <u>requirements</u>.

2) For example, <u>wedding dresses</u> can be <u>custom-made</u> to perfectly fit the bride, and <u>made-to-measure furniture</u> is often produced to <u>fit</u> the dimensions of a <u>room</u>.

3) The <u>workforce</u> needs to be <u>highly skilled</u>, especially if the design is quite detailed — so it's an <u>expensive</u> way to make things.

4) This type of production is also very <u>labour-intensive</u> — it takes a lot of <u>time</u> to make each product.

5) <u>Prototypes</u> for a new product are normally produced as a <u>one-off</u>. If the prototype works well, the product might then be manufactured in <u>greater volumes</u>...

Prototypes are full-size working products or systems. They're used to test a product against its design specification (see p.114 for more on this), so they only need to be made in small volumes.

Batch Production Makes a Certain Number of Products

1) Batch production is where a <u>specific quantity</u> (a <u>batch</u>) of a <u>product</u> is made. Batches can be <u>repeated</u> as many times as necessary.

2) You do <u>one process</u> (e.g. cutting out) on <u>the whole batch</u>, then do <u>another process</u> (e.g. painting the parts you cut out). So it's <u>quicker</u> than making <u>one-off</u> products over and over again.

3) Batch production can be used to manufacture a <u>load of one product</u> (sofas, say), then a load of something <u>a bit different</u> (e.g. armchairs). <u>Printed circuit boards</u> (PCBs — see p.24) are often batch produced as different electronic products require <u>different PCB designs</u>.

Joe's flexibility impressed the interview panel

4) <u>Templates</u>, <u>jigs</u> and <u>moulds</u> (see p.42-43) are often used to make sure products <u>within a batch</u> are <u>identical</u>. A design with <u>less detail</u> (than that typically found in <u>one-off production</u>) can help to achieve <u>consistency</u> too.

5) The <u>machinery</u> and <u>workforce</u> used need to be <u>flexible</u>, so they can quickly <u>change</u> to making a batch of a <u>different product</u>.

6) The time <u>between</u> batches, when <u>machines</u> and <u>tools</u> may have to be <u>set up differently</u> or <u>changed</u>, is called <u>down time</u>. This <u>wastes money</u> because you're <u>not making</u> anything you can <u>sell</u>.

Der der der der, derder-der der der — it's batch of the day...

When revising the differences between the types of production, it's often useful to have a few examples of products that are made by each one. This can really help to jog your memory.

Scales of Production

Mass Production is Making Loads of the Same Product

1) This is the method you'd use to make <u>thousands</u> of <u>identical products</u>, like <u>newspapers</u>, <u>magazines</u> and <u>cars</u>. You'd only use this for a <u>mass-market product</u> — where loads of people want the same thing.

2) The different <u>stages</u> of production are <u>broken down</u> into simple <u>repetitive tasks</u>. Each worker only does a <u>small part</u> of the process, and the product moves further down an <u>assembly line</u> for each stage. Products should have <u>pretty basic designs</u> to make the assembly line simpler.

3) Mass production often uses <u>expensive specialised equipment</u> and <u>CAD/CAM</u> (see p.4).

4) <u>Recruitment</u> is relatively <u>easy</u> — most of your staff <u>don't</u> need to be highly <u>skilled</u>. <u>Robots</u> are increasingly used on assembly lines too (see p.2).

Continuous Production is Non-Stop

Both mass production and continuous production can keep manufacturing costs fairly low — this is partly because the initial cost of equipment is spread over a large number of products.

1) Continuous production <u>differs</u> from mass production in that it <u>runs all the time</u>, without interruption, <u>24 hours a day</u>. That's because it would be too <u>expensive</u> to keep <u>stopping and restarting</u> the process.

2) It's pretty much entirely <u>automated</u> — very <u>few workers</u> are needed (thankfully)...

3) The equipment is built to make <u>huge amounts</u> of only <u>one thing</u>. This means it's expensive but it can be designed to be <u>very efficient</u> — so the <u>cost per item</u> is <u>cheap</u>.

4) Continuous production is used to make products such as <u>aluminium foil</u> and <u>chemicals</u>.

Continuous production — feels more like continuous revision...

So, products can be made as a unique one-off, in a group called a batch, in really massive numbers or in a continuous, never-ending, keep-going-until-the-end-of-time way. And with keeping going in mind...

Q1   Put one-off, batch and mass production in order of scale (from smallest to largest).   [1 mark]

Q2   What type of production is normally used to make a product prototype?   [1 mark]

Q3   How is batch production carried out?   [1 mark]

Q4   Explain what is meant by 'down time' in batch production.   [2 marks]

Q5   Suggest an example of a product that is likely to be mass produced.   [1 mark]

Q6   A company wants to manufacture 100 000 identical nappies.

    a) What type of production would be best suited for this?   [1 mark]

    b) What is the advantage of this type of production when it comes to hiring workers?   [1 mark]

Q7   Give one reason why continuous production runs non-stop.   [1 mark]

Quality Control

You should always <u>check</u> a product <u>works</u> and <u>looks right</u> before <u>selling</u> it — this is known as <u>quality control</u>.

Quality Control Checks the Quality of Manufacturing

1) <u>Quality control</u> ensures products are <u>manufactured</u> to a <u>high enough standard</u>. Checks should be made at <u>every stage</u> of the <u>manufacturing process</u>.

2) This involves <u>testing</u> materials, products or components to check they've been made to a <u>high enough standard</u> and they <u>meet</u> the <u>manufacturing specification</u>. E.g. a musical greetings card might be tested to make sure all the images are <u>printed clearly</u>, the edges are <u>cut accurately</u>, the card is <u>folded</u> in the <u>right place</u>, the text is <u>straight</u>, it plays the <u>right tune</u>, etc.

3) Usually, only a <u>sample</u> of the materials, components or products are <u>tested</u>, as it would take <u>too much time</u> to check them <u>all</u>.

Manufacturing specifications (see p.112-113) include manufacturing instructions and exact details of what the product should be like.

Tolerance is the Required Accuracy of a Measurement

1) It's important that the <u>dimensions</u> (size) of components are <u>accurate</u>. If they aren't, the parts won't <u>fit together</u> properly when the product is assembled.

2) Just <u>how accurate</u> they need to be is <u>specified</u> by the <u>tolerance</u>. This is the <u>margin of error</u> that is considered to be <u>acceptable</u>, e.g. small enough <u>not</u> to <u>affect</u> the product's <u>functionality</u>.

3) The tolerance is normally given as an <u>upper</u> (+) and <u>lower</u> (–) <u>limit</u> for the measurement. For example, a <u>20 mm</u> measurement with a tolerance of <u>±0.5 mm</u> has a lower limit of <u>19.5 mm</u> and an upper limit of <u>20.5 mm</u>.

Tolerance can be applied to loads of different types of measurement — not just size. For example, measurements of physical properties (e.g. weight), chemical properties (e.g. the composition of an alloy) or electrical properties (e.g. the resistance of a resistor, see p.88).

EXAMPLE: A shelf is designed to be 200 mm wide by 720 mm long, with a tolerance of ± 5.0 mm. Would a 203 mm by 728 mm shelf be within this tolerance?
The width should be 200 ± 5.0 mm — between 195 mm and 205 mm, which it is.
The length should be between 715 mm and 725 mm, which it isn't, so it's not within the tolerance.

4) Tolerances should be included on <u>working drawings</u> (in the <u>manufacturing specification</u>) to show the <u>limits</u> within which the product should be manufactured.

5) Components must be <u>within the tolerance</u> to <u>pass</u> through any <u>size checks</u> in quality control.

There are Many Different Quality Control Tests

Go/No Go Fixtures Check Dimensions are Within Tolerance

1) Go/no go fixtures are <u>limit gauges</u> — they <u>check</u> to see whether the size of a part is <u>within its tolerance</u>.

2) They're usually <u>double-ended</u> — one end is machined to the <u>lower limit</u> and the other end to the <u>upper limit</u> of tolerance.

3) These checks <u>don't take very long</u> and are <u>much quicker</u> than <u>measuring</u> the <u>actual dimensions</u> of a component, e.g. using a ruler or micrometer.

4) They're often used in the quality control of <u>timber-based</u> components.

If the component <u>fits through</u> the <u>upper limit</u> (go) but <u>not</u> the <u>lower one</u> (no go), it's <u>within the tolerance</u>.

Registration Marks Check Printing Plates are Aligned

1) A <u>colour registration mark</u> normally appears as a <u>cross shape</u> — it is used by manufacturers to <u>check</u> the <u>quality</u> of <u>colour printing</u> onto <u>paper</u> and <u>board</u>.

2) They're used to make sure the <u>printing plates</u> are <u>aligned</u> in the right position.

Colour printers use four printing plates — cyan, magenta, yellow and black (see p.55).

If the plates <u>are</u> all in the right places you get a single, <u>clear image</u>..

... and if not, the <u>registration mark</u> will be printed a bit <u>fuzzy</u>.

Quality Control

Prints Should be Checked Against the Original

1) To check <u>repeating prints</u> (such as stripes and chequers) are being <u>printed</u> correctly onto <u>fabrics</u>, manufacturers will often <u>compare prints</u> to an <u>original sample print</u>. This can be done pretty well <u>by eye</u>.

2) It's much <u>quicker</u> and <u>easier</u> to do this than try to <u>measure</u> the <u>dimensions</u> of each part of the repeating pattern <u>one at a time</u>.

There are Ways to Achieve Consistency During Manufacture

Depth Stops Control the Depth of Drilling and Cutting

1) Depth stops are <u>long rods</u> that are <u>clamped</u> close to the drill bit of <u>some drills</u>.

2) They allow you to <u>drill</u> a hole to an <u>exact depth</u> in whatever material you're drilling. Once this depth has been <u>reached</u>, the depth stop will come into <u>contact</u> with the <u>material</u> and will <u>prevent</u> you from <u>drilling</u> any <u>deeper</u>.

3) The drill depth can be <u>adjusted</u> by <u>clamping</u> the depth stop in a <u>different place</u>.

4) Along with the <u>width</u> of the drill bit, this allows you to carefully <u>control</u> the <u>dimensions</u> of the <u>hole</u>.

Depth stop

Depth stops can also be used with certain types of saw, allowing you to cut to a certain depth but no further.

Laser Cutters Need to be Programmed Correctly

1) To <u>cut</u> a material <u>accurately</u>, a laser cutter needs to be <u>programmed</u> with the <u>right information</u>. This includes:

There's more on laser cutters on page 5.

- The <u>dimensions</u> of the <u>component</u> to be cut — so it knows <u>exactly where to cut</u>.
- The <u>correct power settings</u> and <u>feed rate</u> (the <u>speed</u> that the <u>laser moves</u> over and cuts the material). These values depend on the <u>type</u> of material to be cut and its <u>thickness</u>.

2) If <u>programmed correctly</u> the laser can cut to a <u>tiny tolerance</u> (with <u>high accuracy</u> and <u>precision</u>), so the component should end up <u>as you want it</u>.

3) A <u>sample</u> still needs to be <u>checked</u> though to make sure the <u>machine</u> hasn't <u>malfunctioned</u> or been <u>incorrectly programmed</u>.

PCBs Should be Made with the Same Exposure Times

1) <u>Printed circuit boards</u> (PCBs) can be manufactured through <u>photo-etching</u>.

2) In this process, the PCB is <u>exposed</u> to <u>UV light</u>, a <u>developer solution</u> and an <u>etching solution</u> — this <u>removes unwanted copper</u> from the board, <u>leaving behind</u> the copper needed to form the <u>tracks</u>.

3) <u>Exposure times</u> have to be <u>carefully selected</u>. For example, <u>exposing</u> the PCB for <u>too long</u> will cause the <u>copper tracks</u> to be <u>removed</u>, but <u>too little time</u> will mean <u>unwanted copper remains</u> on the board.

Photo-etching is covered on page 92.

4) Once chosen, exposure times should be <u>kept the same</u> for <u>every PCB</u> that's produced. This helps to make sure that the PCBs are all of a <u>consistent quality</u> when output.

I can assure you — this page is quality...

Q1 A board needs to be cut with dimensions of 135 × 180 mm with a tolerance of ± 3 mm. Which of the following falls within the stated tolerance?

A: 38.1 × 177.5 mm   B: 134.4 × 183.2 mm
C: 133.9 × 181.8 mm   D: 138.0 × 176.7 mm   [1 mark]

Q2 What is a go/no go fixture used for?   [1 mark]

Q3 What piece of equipment should be used to control the depth of a hole being drilled?   [1 mark]

Production Aids

Production aids do exactly what they say on the tin — they're tools and techniques used to aid production. They often speed up or simplify a process, or are used to help control accuracy...

All Points Can be Defined Relative to a Reference Point

1) A reference point is a point where measurements are made from — this could be the corner of an object. You can also have reference lines (e.g. the x-axis, or an object's edge) and reference surfaces that you measure from.

2) Using reference points, lines and surfaces helps to control accuracy, e.g.

> Imagine you want to make a coat rack with a coat hook every 10 cm. You wouldn't measure each hook 10 cm from the previous one, as if the previous hook is slightly off, the inaccuracy would be carried over. Instead, you'd mark out where each hook will go by measuring each one from a single reference point, e.g. at one end of the coat rack — this is quicker and reduces measurement error.

3) When x,y,z coordinates are used, the reference point is often the datum — the point where x, y and z meet (0,0,0).

4) Reference points are really important in aligning CAM machines with the materials they're operating on. CAM machines also use x,y,z coordinates to guide the tools being used (see p.4).

5) Reference points are useful when scaling the size of something too, e.g. a drawing:

- First, choose a reference point, e.g. one corner of the drawing.
- Then, figure out how far away each point on your drawing is from the reference point. This is basically their (x,y) coordinates if the reference point is the datum, i.e. (0,0).
- Next, multiply these coordinates by a scale factor — for example, if you wanted to make the drawing twice as big, you'd multiply by two, e.g. (0,4) would become (0,8).
- Finally plot your new coordinates and ta-da, you've scaled your drawing.

Reference point

> Scaling can be really useful. For example, you might want to scale a pattern (see next page) to produce a piece of clothing in a different size.

Templates are Used for Drawing and Cutting Round

1) Templates are very easy to make and simple to use.

2) You can draw round or cut round them to produce a shape that is identical to the template.

3) This is much quicker and easier than measuring out the dimensions of the shape each time.

4) As long as your template is accurate the batch of components you make with it should all be accurate too — using a template makes the products consistent.

Template

5) Templates need to be strong and hard-wearing — so that they can be used repetitively without getting damaged or worn.

> If you're cutting lots of shapes from a sheet of material, you need to plan how you want to lay the template out — some layouts will result in less wastage of material (e.g. see p.117).

6) They're useful in quality control too — you can check your components against the template to make sure they're the same.

The boy band template — strong, hard-wearing and definitely used repetitively

Production Aids

Patterns are Used with Textiles and for Casting

1) In textiles, patterns are <u>templates</u> that are used to <u>cut out pieces</u> of <u>fabric</u> — these <u>pieces</u> are then <u>sewn together</u> to make a <u>textiles product</u>.

2) <u>Patterns</u> are normally made from <u>tissue paper</u> — they're <u>pinned</u> to a fabric so that you can <u>cut round</u> them.

3) In <u>industry</u>, <u>CAM cutting machines</u> are used to <u>cut</u> the fabric. The <u>patterns</u> are drawn out using <u>CAD</u>. The <u>coordinates</u> of the <u>pattern's outline</u> are then used to tell the machine <u>how to cut</u> the pieces of fabric.

Head hole

Arm hole

Example of a vest pattern — pieces of fabric cut using this could be sewn together to form a vest.

1) <u>Casting</u> involves forming a material into a <u>particular shape</u> — see p.68-69. This shape is determined by a <u>pattern</u> — which is an <u>exact replica</u> of the object you want to form.

2) Patterns can be made from materials such as <u>resin</u>, <u>wood</u> and <u>metal</u>.

3) Patterns can be <u>used many times</u> to make moulds that are used in the casting process and make products that are a <u>consistent shape</u>.

Example of a metal part that has been cast using a pattern.

Jigs Guide Tools to Simplify and Speed Up Production

1) A jig <u>guides</u> the <u>tools</u> that are working on a component, or makes sure that the <u>workpiece</u> is <u>positioned</u> in the <u>right place</u> for machining.

2) <u>Jigs</u> come in many <u>different shapes and sizes</u> and can be <u>specifically</u> made for a particular job.

3) They're designed to <u>speed up production</u> and <u>simplify</u> the <u>making</u> process.

Although jigs are really useful, they can take a long time to make. So it's not worthwhile using one if you're only going to make a few products.

metal guide for holes

drilling jig

finished product

Using a <u>drilling jig</u> means you <u>don't</u> have to <u>mark out</u> exactly <u>where</u> you want <u>to drill</u> — this <u>saves time</u> and <u>effort</u>. It <u>guides</u> the drill, which helps to cut down on <u>errors</u>, and makes sure that components are <u>consistent</u>.

A <u>dovetail jig</u> allows complex, <u>dovetail joints</u> to be cut <u>quickly</u>, <u>easily</u>, and with <u>minimal measuring</u> and marking out.

Dovetail jig

Dovetail joints

No — there's nothing on this page about Scottish dancing...

In short, reference points, lines and surfaces, templates, patterns and jigs make life easier. Sadly, you'll need to write a bit more than that if they come up in the exam — handily, all the details you need are on these pages.

Q1 a) What is a reference point? [1 mark]

b) Give two ways in which reference points aid production. [2 marks]

c) What is commonly used as the reference point when working with x,y,z coordinates? [1 mark]

Q2 What are patterns for textiles products normally made from?

A: fabric   **B**: tissue paper   **C**: board   **D**: tracing paper [1 mark]

Q3 a) Give one advantage of using a jig when making components. [1 mark]

b) Adam is making a one-off product. Explain why using a jig won't save him time overall. [1 mark]

Production of Materials

New materials need to be extracted from a raw material source and processed into a useful form.

Paper and Board are Made from Cellulose Fibres

1) Trees go through a series of processes before they become paper or board:

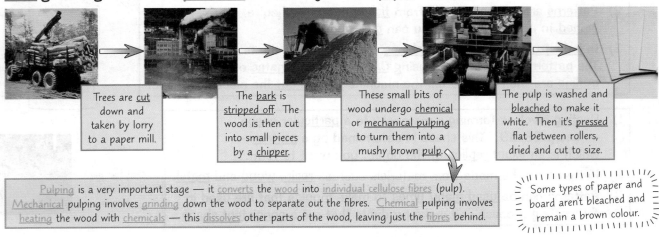

Trees are cut down and taken by lorry to a paper mill.

The bark is stripped off. The wood is then cut into small pieces by a chipper.

These small bits of wood undergo chemical or mechanical pulping to turn them into a mushy brown pulp.

The pulp is washed and bleached to make it white. Then it's pressed flat between rollers, dried and cut to size.

Pulping is a very important stage — it converts the wood into individual cellulose fibres (pulp). Mechanical pulping involves grinding down the wood to separate out the fibres. Chemical pulping involves heating the wood with chemicals — this dissolves other parts of the wood, leaving just the fibres behind.

Some types of paper and board aren't bleached and remain a brown colour.

2) The cellulose fibres used for paper and board can also come from other plants like grasses. The process is similar to the conversion of wood to paper, although no debarking or chipping is required.

3) Certain types of board may have an extra stage in production. For example, corrugated cardboard will need its fluted middle layer sticking to the outer layers to form a laminate.

Wood is also Cut Down for Timber

1) Trees, e.g. from a plantation or a forest, are felled (cut down).

2) The bark is removed, and the trunk is sawn up. Which way it's sawn affects how the planks will look, and how much they're likely to bend or warp (twist).

Slab sawn   Quarter sawn

3) The wood is then seasoned by drying it...

Wood can be 'air-dried' or 'kiln-dried':

- Air-drying involves stacking wood up outside or in sheds and allowing air to circulate around it.

- Kiln-drying is similar to air-drying but it's done in a heated chamber with fans to circulate the air. This makes it loads quicker, but the conditions have to be carefully controlled to avoid the wood drying too fast (which can damage it).

Seasoning makes the wood stronger and less likely to rot (see p.59).

4) Seasoned wood can be converted into lots of different useful forms, e.g. planed square edge wood, mouldings, etc. There's more about the forms wood is available in on p.60.

5) Wood can also be processed into manufactured timbers (see p.23):

- Medium density fibreboard, MDF uses wood that has been processed into cellulose fibres, in a similar way to paper and board production (e.g. debarked, chipped, and pulped).

- Chipboard uses a mixture of dried wood chips, shavings and sawdust — this is often material that has been discarded as waste from another process.

Glue (e.g. urea formaldehyde) is added to these mixtures.
They are heated, pressed into sheets and left to dry.

Plywood uses wood that has been softened, either through soaking it in hot water or steaming it. A thin sheet is peeled from the softened wood, cut to suitable size and dried. The cut wood can then be arranged into stacks of three or more layers. Each layer has a grain direction that is at 90° to the layer above and below (see p.37).
Glue is added between each layer and the sheets are heated and pressed.

Once they're dry, manufactured boards can be trimmed and sanded to give them a smooth finish.

Production of Materials

Metals Come from Rocks in the Ground

1) Metal is <u>mined</u> from the <u>ground</u> as a <u>metal ore</u> — a rock with enough metal locked up in it to make it <u>profitable</u> for the metal to be <u>extracted</u> (separated) from the rest of the ore.

2) <u>Extraction</u> can be achieved in <u>two</u> main ways — through <u>heating</u> the ore in a <u>furnace</u> or by <u>electrolysis</u>:

> Some metal ores can be <u>crushed</u> and then <u>heated</u> in a <u>furnace</u> with <u>other materials</u> such as <u>charcoal</u>. The metal <u>separates</u> out and will often <u>sink</u> to the bottom where it can be <u>tapped off</u>.

> <u>Electrolysis</u> uses <u>electricity</u> to extract a metal from its ore.

Sometimes the ores need to be <u>prepared</u> before extraction can take place. For example, <u>bauxite</u> (see below) is <u>heated under pressure</u> with <u>chemicals</u> to <u>separate aluminium oxide</u> from the mixture. This then undergoes <u>electrolysis</u>.

3) Metals that have been extracted from their ores will <u>contain impurities</u> — <u>small amounts</u> of other <u>substances</u>. These impurities will often need to be <u>removed</u> before the metal can be <u>used</u> in a product. This process is known as <u>refining</u>, and there are several ways of achieving this (see below).

4) The refined, <u>molten</u> metal is usually <u>cast</u> — this process involves pouring the metal into a <u>mould</u> and allowing it to <u>cool</u> and <u>solidify</u> into the <u>required shape</u>.

5) Here's a bit more on the processes involved in the <u>production</u> of a few different metals...

Metal	Ore	Extraction method	Refining method
Iron	E.g. haematite	Furnace	Main impurity is sand — removed by adding limestone to the furnace during extraction.
Aluminium	E.g. bauxite	Electrolysis	Gases can be pumped through molten aluminium. Causes impurities to rise to the surface.
Zinc	E.g. sphalerite	Furnace or electrolysis	Zinc is heated and then cooled. This separates it from impurities with different boiling points.
Copper	E.g. chalcopyrite	Furnace	Electrolysis can be used to separate the copper from any impurities left in it.
Tin	E.g. cassiterite	Furnace	Heating the tin can cause some impurities to react and form a substance that can easily be removed. It can also be used to separate impurities based on melting points. Electrolysis can be used to refine the tin too.

Most Plastics are Made from Oil

Most plastics are made from <u>crude oil</u>, although some can be made from substances found in <u>plants</u>.

1) <u>Crude oil</u> is <u>extracted</u> from the <u>ground</u> (on land or at sea) and taken to a <u>refinery</u>.

2) Here, it is <u>heated</u> using a process called <u>fractional distillation</u>. This <u>separates</u> it into different chemicals called '<u>fractions</u>'.

3) <u>Some</u> of the fractions can be <u>linked together</u> to make <u>polymers</u> (<u>plastics</u>):

> - The chemicals that get joined together are called <u>monomers</u>.
> - The monomers join up to make <u>chains</u> called <u>polymers</u>.
> - So the process is called <u>polymerisation</u>.

Monomers   Polymerisation   Polymer

4) <u>Other fractions</u> may need to be <u>broken down</u> into <u>smaller molecules</u>. This is done by <u>heating</u> these fractions in a process called <u>cracking</u>. These chemicals can then be <u>polymerised</u>.

5) The plastics can then be <u>used</u>. For example, they are often <u>cut up</u> into <u>small pieces</u> and <u>moulded</u> into the <u>right shape</u> (see pages 69-71 for more on how plastics can be moulded).

I'm feeling chipper — revision is such a blast...

Make sure you know the steps involved in getting each material into a useful form. Try scribbling them down now.

Q1   a) Which raw material are plastics usually made from?   [1 mark]

       b) What is fractional distillation?   [1 mark]

More on the Production of Materials

Fabrics are made up of fibres. There's a huge range of fibres that are used in the textiles industry.

There are Many Types of Fibres

1) Fibres are tiny 'hairs' that are spun into yarns — the threads that are woven or knitted into fabrics.

2) Fibres come in either short lengths (staple fibres), or longer lengths (filaments). Filaments can be cut up into staple lengths if required.

3) Fibres fall into three main categories — natural, regenerated and synthetic fibres...

Natural Fibres are Harvested from Natural Sources

1) Natural fibres are fibres obtained from natural sources (e.g. plants and animals).

2) First, they're harvested and processed, for example...

WOOL is from sheep's fleece

| The sheep is sheared — the best wool comes from the sheep's shoulders and sides. | ⇒ | Cleaning and scouring (washing with harsh chemicals) removes grease and dried sweat. | ⇒ | The fibres are then combed using wire rollers — this is called carding. |

COTTON is from the seed pods of the cotton plant

| Before harvesting, the plants are treated with chemicals to make the leaves fall off. | ⇒ | Fibres are cleaned to remove dirt. | ⇒ | Seeds are removed. | ⇒ | Carding (see above). |

SILK comes from the cocoon made by silk worms

| The cocoon protects the worm whilst it turns into a moth. | ⇒ | The cocoon is one long piece (a filament) of silk fibre. It's stuck together with a hard gum (called sericin). | ⇒ | The gum is softened by soaking in warm water. The filament is then unwound (this is called reeling). |

cocoon

3) The fibres can then be spun into yarn. These are then made into fabric — see p.21.

Regenerated Fibres are Chemically Treated Natural Materials

1) Regenerated fibres are made from natural materials — usually cellulose from wood pulp. However, these materials are chemically treated to produce fibres.

2) Different fibres are made by using different chemicals — e.g:

Viscose fibres are made by dissolving cellulose in sodium hydroxide solution. The liquid is forced through tiny holes, and hardened to form filament fibres. These filaments are stretched into a yarn which is wound onto spools, or chopped into staple lengths.

You can think of regenerated fibres as somewhere between natural and synthetic fibres — they are semi man-made.

Synthetic Fibres are Made from Polymers

Synthetic fibres are man-made fibres. They're made from polymers — long chains of monomers that come mainly from oil. For example, polyester, LYCRA® and nylon are produced from crude oil...

Different polymers will make different synthetic fibres.

| Crude oil is fractionally distilled. Some of the products are made into polymers (see previous page). | → | The polymers are melted and forced through tiny holes. | → | They're then cooled to form long filaments. |

The filaments are made into yarn. This is wound onto spools, or chopped into staple lengths to be spun.

More on the Production of Materials

Material Production Can Impact the Environment

1) The <u>conversion</u> of <u>raw materials</u> into useable forms can have a <u>negative effect</u> on the <u>environment</u> — it can contribute to <u>pollution</u>, the <u>destruction</u> of <u>natural habitats</u> and <u>climate change</u>. (There's more on this on p.8 and p.35.)

2) Just <u>obtaining</u> the <u>raw materials</u> causes environmental problems. For example:

Obtaining wood for <u>timber</u>, <u>paper</u> and <u>board</u> can cause <u>deforestation</u>. This is where large areas of <u>forest</u> are <u>cut down</u>, and trees <u>aren't planted</u> to <u>replace</u> the old ones. This <u>destroys</u> the <u>forest habitat</u>, which has a negative effect on the <u>animals</u> and <u>plants</u> that live there.

Sustainably managed forests (where trees are replanted) are a much more environmentally friendly source of wood.

<u>Drilling</u> for <u>oil</u> is the <u>first step</u> in the production of <u>polymers</u> and <u>synthetic fabrics</u>. Oil can be drilled both on <u>land</u> and <u>out at sea</u>. Drilling can release <u>toxic</u> chemicals into the <u>atmosphere</u>. Sludgy <u>waste material</u> or <u>oil</u> can also <u>leak</u> and <u>pollute</u> surrounding land or marine <u>habitats</u> — this can be <u>harmful to the wildlife</u> living there. On land, habitats need to be <u>cleared</u> to <u>make room</u> for the <u>drill sites</u>.

<u>Mining metals</u> uses a lot of <u>energy</u> from <u>fossil fuels</u> — this causes <u>air pollution</u> and contributes to <u>global warming</u>.

<u>Habitats</u> are <u>destroyed</u> when large areas of land are <u>cleared</u> for <u>mines</u>. The <u>surrounding</u> habitats can be affected too — <u>chemicals</u> (used in mining) and <u>waste rock</u> can <u>pollute</u> nearby <u>water</u>, which can be <u>harmful</u> to <u>wildlife</u>.

Oil and metals are both examples of finite resources — once they've been used up there won't be any more (see page 6 for more about this).

Pesticides are chemicals used to kill weeds and insects that harm the crops. Fertilisers are chemicals that help crops to grow.

<u>Farming natural fibres</u> (e.g. cotton) often uses <u>artificial fertilisers</u> and <u>pesticides</u> to help increase the amount of natural fibre produced — these can <u>pollute rivers</u> and <u>harm wildlife</u>. Farming also needs lots of land — getting it may involve <u>deforestation</u> or <u>clearing habitats</u>.

Cotton can harm the environment — I'm going commando...

This all sounds a bit gloomy, doesn't it. There is hope though — we can help the environment by reusing and recycling the materials we've already produced. That way, we don't have to produce as much new stuff, reducing the damage to the environment.

Q1   a) Which of the following is an example of a regenerated fibre?

      A: cotton   **B**: silk   **C**: viscose   **D**: polyester   [1 mark]

    b) Name two synthetic fibres.   [2 marks]

Q2   Give two ways in which farming natural fibres can harm the environment.   [2 marks]

Properties of Paper and Board

Each type of paper and board has properties that make it more suited for certain tasks. Let's have a look at some of the physical properties that determine how they're used in different products...

Different Properties are Useful for Different Things

1) Different sorts of paper and board have pros and cons — you need to be ready to weigh them up.

Flexibility and rigidity

To make 3D products, you need paper and board that can be bent or folded without breaking but is rigid enough to keep its shape (if it's used as a box, say). Corrugated board is a good option when you need quite a strong, stiff material.

Strength and weight

1) Some materials (like corrugated card) can withstand a fair amount of force without breaking — good for heavy duty packaging or products that will be handled a lot.

2) But stronger materials are usually bulkier which can impact transport costs — the more bulky they are, the more they cost to transport.

Toxicity and sustainability

1) Recycled paper and board may contain toxic chemicals which mean they're not suitable for use as food packaging.

2) Laminated paper and board (see p.56) can be hard to recycle because it's hard to separate the paper/board from the other materials.

Cost and quality

1) Expensive paper and board makes a product feel high-quality.

2) But it is only worth using for luxury products or something that has to last a long time.

Finish

1) Solid white board has an excellent surface for printing on.

2) A cheaper alternative is to use recycled board laminated with high quality paper.

3) Plain recycled board can be used for something that doesn't need a good finish.

2) When choosing a material, you need to consider its use. Here are some examples:

Card based food packaging

1) Cardboard food packaging needs a combination of properties. It needs to be printable, so that the customer can see the advertising and nutritional information for what they are buying. But it often also has to be waterproof and airtight, to prevent the food inside from going bad.

2) A combination of aluminium foil and board can be used to package food, e.g. soup cartons. It keeps flavours in and air out. You can also print graphics onto the paper. Duplex board (look back to p.16) is often used for food packaging because it is strong and easy to print on.

3) Pizza boxes are made of corrugated cardboard that is strong enough to withstand other boxes being stacked on top of it. It's thick, so it's good at retaining heat, which keeps the pizza warm.

Flyers and leaflets

1) Flyers and leaflets need to be produced cheaply and often in large quantities. Usually the quality of print doesn't need to be high as it doesn't need to last for a long time. For these reasons, they're often printed on cheap, low-weight (low gsm) paper.

gsm stands for grams per square metre.

2) Some leaflets for classier products might be printed on heavier, higher quality paper because it suits a higher quality product.

3) Biodegradable paper would work well, as the leaflets are often dropped on the street or thrown away.

4) Unbleached paper may also be a good option as it is more environmentally friendly.

If you laminate paper by adding a layer of another material, you get a composite (see p.33) with different properties.

Properties of Paper and Board

Properties of Paper and Board can be Changed

1) Paper and board can be <u>modified</u> for a specific purpose. <u>Chemicals</u> called additives can be added <u>during manufacture</u>, or can be used to coat paper and board to give it desirable properties.

2) For example, <u>chemical additives</u> can increase <u>paper strength</u>, enhance <u>brightness</u> or add <u>colour</u>.

3) Another example is to <u>prevent moisture transfer</u>, i.e. moisture <u>passing through</u> the paper or board.

> ### Examples
>
> 1) <u>Kraft paper</u> undergoes a chemical treatment that makes it <u>strong and tear-resistant</u>. It's also a <u>vapour barrier</u>, so can be used on food packaging for things like flour, which needs to <u>stay dry</u>.
>
> 2) <u>Baking parchment</u> is usually treated with sulphuric acid to give it its <u>non-stick properties</u>.
>
> 3) <u>Greaseproof paper</u> is treated to change the qualities of its fibres — making it <u>less absorbent</u>.
>
> 4) <u>White board</u> is a <u>strong</u>, <u>thick</u> board that is <u>good for printing on</u>.
> It's the only type of board recommended for <u>direct contact</u> with food.
> The board is <u>bleached</u> and then coated with <u>wax</u> (or laminated with polyethylene).
>
> 5) The pulp used to make <u>paper towels</u> is treated with <u>different resins</u> during manufacture to increase its <u>wet strength</u>. This makes sure that the towels <u>don't break</u> when they become <u>wet</u>.

Paper and Board Come in Different Shapes and Sizes

1) There are <u>various types</u> of paper and board available (take a look back to p.16).

2) Paper and board can be bought by the <u>sheet</u> and in <u>rolls</u> — these are known as <u>stock forms</u> (also see p.60).

3) Sheets of paper and board are sold in <u>standard sizes</u>. Sizes go <u>from A0</u> (which has an area of 1 m²) to <u>A1</u>, <u>A2</u>, and so on — halving in size (area) each time.

4) Many other sizes are also available — <u>A4</u> is <u>half</u> the size of <u>A3</u>, <u>A5</u> is <u>half</u> the size of <u>A4</u>, <u>A6</u> is <u>half</u> the size of <u>A5</u>, etc.

5) As the size gets smaller the number <u>increases</u>.

6) The most common paper sizes used in schools are <u>A4</u> and <u>A3</u>. The <u>width</u> of A3 paper is the <u>length</u> of A4. The <u>length</u> of A3 paper is <u>double the width</u> of A4.

7) Paper and board are also available in different <u>weights</u> (see p.16), <u>thicknesses</u> and <u>colours</u>.

8) You can choose the <u>ply</u> of your material too. Ply means how many <u>layers</u> it is made out of — 1-ply is one layer, 2-ply is two layers and so on.

A roll of corrugated board

I'm board by this page — bet that was a joke you didn't see coming...

When you're picking a type of paper or board for a project you need to think about what you need it to do, and select the right properties. Don't forget that you can change the properties too...

Q1	Why might recycled paper or board not be suitable for use in food packaging?	[1 mark]
Q2	Give two properties that are useful for food packaging to have.	[2 marks]
Q3	Why are additives added to paper and board?	[1 mark]
Q4	How are paper and board commonly sold?	[1 mark]
Q5	How much larger than A4 is A2?	[1 mark]

Standard Components

Standard components are parts that come <u>ready-made</u> for use. Here's two pages filled with them...

Standard Components are Pre-Manufactured Parts

1) Standard components are common fixings and parts that <u>manufacturers buy</u> instead of manufacturing them <u>themselves</u>, e.g. screws, rivets and buttons.

2) Standard components are <u>mass produced</u> so they're available at <u>low cost</u> to manufacturers.

3) Using standard components saves <u>time</u> during manufacture — so it is more <u>efficient</u>.

4) <u>Specialist machinery</u> and <u>extra materials</u> aren't needed so it also saves <u>money</u>.

5) So, when <u>designers</u> are working on a new product, they need to think about which standard components they could use. When using CAD, designers can <u>select</u> standard components to use in their designs.

Lots of Standard Components are Used with Paper and Board

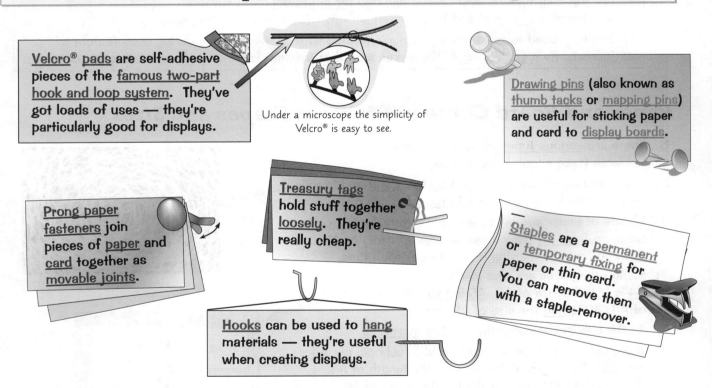

<u>Velcro®</u> <u>pads</u> are self-adhesive pieces of the <u>famous two-part</u> <u>hook and loop system</u>. They've got loads of uses — they're particularly good for displays.

Under a microscope the simplicity of Velcro® is easy to see.

<u>Drawing pins</u> (also known as <u>thumb tacks</u> or <u>mapping pins</u>) are useful for sticking paper and card to <u>display boards</u>.

<u>Prong paper</u> <u>fasteners</u> join pieces of <u>paper</u> and <u>card</u> together as <u>movable joints</u>.

<u>Treasury tags</u> hold stuff together <u>loosely</u>. They're really cheap.

<u>Staples</u> are a <u>permanent</u> or <u>temporary fixing</u> for paper or thin card. You can remove them with a staple-remover.

<u>Hooks</u> can be used to <u>hang</u> materials — they're useful when creating displays.

Bindings use Standard Components

Comb binding

1) <u>Holes</u> are punched in sheets using a special machine, then a <u>plastic comb</u> is inserted.

2) Pages can be added or removed <u>without causing damage</u>.

3) The book opens flat — <u>easy to read</u>.

Spiral binding

1) A <u>plastic coil</u> is inserted down the spine.

2) <u>Wiro binding</u> is similar but uses a <u>double loop wire</u> instead of a plastic coil.

Comb binding

Spiral binding

Wiro binding

Saddle stitching

1) <u>Double-sized pages</u> are folded and <u>stapled together</u> at the centre.

2) Easy and cheap — but it <u>won't</u> <u>hold many sheets</u>.

3) The books open more or less <u>flat</u>.

Standard Components

Some Bindings are More Expensive

Perfect binding

1) Pages are folded together in <u>sections</u>.
2) Each section is <u>roughened</u> at the fold and then <u>glued</u> to the <u>spine</u>.
3) You can bind <u>lots of sheets</u> but you can't open the book <u>flat</u>.

Thread sewing

1) This method is more <u>expensive</u>.
2) The pages are <u>sewn</u> together in sections, then a <u>soft cover</u> is <u>glued</u> on.
3) The pages are less likely to come <u>loose</u> than with perfect binding.

Case-bound / hard-bound

These books are like <u>thread-sewn</u> ones but with a <u>hard cover</u>.

Seals, Tape and Adhesives Keep Things Attached to Each Other

Envelope seals

1) <u>Self-seal</u> envelopes use two strips of <u>adhesive</u> that seal together on contact.
2) <u>Peel and seal</u> envelopes also have two strips of adhesive but one is <u>covered by a strip</u> (often made of paper). To <u>seal</u> the envelope, just remove the <u>covering strip</u> and <u>press</u> the two strips of adhesive together.
3) These seals are <u>very strong</u> and difficult to break. They are used for important or <u>confidential letters</u> because they can't be opened without tearing the envelope — so you can see if your letter has been <u>tampered with</u>.

Double-sided tape

1) <u>Double-sided tape</u> is useful for attaching two separate pieces of paper or board together <u>without a visible seal</u>.
2) This tape usually has a <u>peelable</u> layer on one side — this is removed when the tape is used.

Tabs

1) <u>Tabs</u> are stickers that can be used to seal something.
2) For example, tabs can be used to keep a leaflet or booklet <u>closed</u> while it <u>travels through the post</u>.

Adhesives

1) Adhesives (<u>glues</u>) can be used to <u>bond</u> paper and card — they can be <u>applied</u> in different ways.
2) <u>Glue sticks</u> uses a <u>solid</u> glue that can be <u>pressed</u> onto a material at a join. It's <u>non-toxic</u>, <u>cheap</u> and <u>environmentally-friendly</u>. Glue can also be applied as a <u>liquid</u>, e.g. using <u>squeezy glue pens</u>. In both cases, the glue is <u>clear</u> when <u>dry</u>.
3) You can also use <u>spray adhesive</u> for <u>mounting photos</u> onto paper or card — they cover large areas well and allow for <u>repositioning</u> (unless it says 'permanent').

Make sure all this information is sealed into your memory...

You probably knew about some of those standard components anyway, or you might have seen them in use. Just make sure you remember some examples of when to use them and why they are useful.

Q1 What are standard components? [1 mark]

Q2 Which of the following components is often used to attach paper or card to a display board?
a) treasury tags   b) drawing pins   c) plastic comb   d) tabs   [1 mark]

Q3 Lauren is choosing a binding for a book. It is a small book without many pages. It needs to be cheap to produce, and must lie flat when opened. Name a suitable type of binding. [1 mark]

Working with Paper and Board

OK, you've probably known about <u>scissors</u> since you were old enough to run with them, but that's no excuse to skip these pages. There are plenty of other, more <u>exotic</u> cutting tools out there.

Scissors can Cut Paper and Thin Card

1) You can cut <u>paper</u> and <u>thin card</u> well with scissors — but they're not much good for <u>fine detail</u> or for <u>removing</u> bits from within a sheet of paper or card (scalpels and knives are good for this — see below).

2) You can use '<u>pinking shears</u>' to produce an interesting <u>zigzag</u> edge (which is commonly used with fabrics — see p.77).

Knives and Cutters also Cut Card and Paper

1) There are loads of different <u>craft</u>/<u>trimming</u>/<u>hobby knives</u>.

2) You mainly use these to cut <u>card</u> and <u>paper</u> — though some will cut thicker board, balsa wood, etc.

Metal cased knife

<u>Stanley</u>® <u>knives</u>, and other 'generic' types. These are good for <u>tougher materials</u>, e.g. thick board or balsa wood.

Scalpels

These are very <u>sharp</u> and great for <u>precision cutting</u>.

Circle cutters

1) You use these to cut <u>arcs</u> and <u>circles</u> in card and paper.

2) You can <u>vary</u> the <u>diameter</u> of the arc or circle to be cut.

Plastic trimming knife

Similar to the metal cased knife, but some have <u>retractable blades</u> or <u>blade covers</u> for <u>safety</u> when not in use.

Perforation cutters

1) These have a round blade which <u>rotates</u> as you push it along, making a line of lots of small <u>cuts</u>.

2) These can be used to make <u>tear-strips</u>, like the ones you find on forms where you're asked to detach a bit and send it back.

These are a bit like pizza cutters.

point   cutting blade

Circle cutters are used like a mathematical compass.

Guillotines and Paper Trimmers Cut Large Sheets

1) <u>Guillotines</u> are used to cut <u>large sheets</u> of paper and card, often <u>many sheets</u> at a time.

2) They have a <u>large blade</u> that you <u>push down</u> to produce a nice <u>straight cut</u>.

3) <u>Paper trimmers</u> are a similar piece of equipment that uses a <u>smaller rotary blade</u> to make straight cuts.

Equipment like guillotines, paper trimmers and die cutters (see next page) use shear forces (see p.36) to cut through paper and board.

Laser Cutters are Expensive but can Make Detailed Cuts

1) <u>Laser cutters</u> are machines that cut out designs drawn using <u>CAD</u>. They can make really <u>accurate</u> and <u>fine cuts</u> through paper and card.

2) They use a very fine laser beam to <u>burn away</u> material. This can be used to make intricately <u>detailed</u> decorative work that can be found on some <u>greetings cards</u> and <u>lampshades</u>. (There's more on laser cutters on p.5.)

Working with Paper and Board

Scoring Makes Paper and Board Easier to Fold

1) If you want to fold something <u>neatly and accurately</u>, then scoring the paper or board first will help.

2) <u>Scoring</u> makes a <u>small indent</u> on the page where it should be folded. To score by hand:

> 1) Draw a <u>guideline</u> with pencil where you want to fold, and line up a <u>metal ruler</u> along it.
> 2) <u>Run a knife</u> (e.g. a trimming knife) along the line, using the <u>ruler for guidance</u>.
> 3) Be careful to only <u>press lightly</u> with the knife to avoid cutting through the paper/board.

3) Scoring makes sure that <u>folds are straight</u> and <u>in the right place</u>.

4) Scoring can also be done using <u>CAD/CAM</u>, with only certain lines in a design being scored.

Die Cutting is Used for Cutting and Creasing Nets

1) Die cutting is used to <u>cut shapes out of</u> and <u>crease lines into</u> paper and board.

2) It's used to produce the nets of complex shaped products like <u>packaging</u>.

3) A <u>net</u> is a <u>2D</u> plan for making a <u>3D</u> object.

4) You can use CAD to design a net, and then use a <u>die cutter machine</u> to cut it out. These can be manual machines, operated by hand, or they can be computerised CAM machines.

5) The die itself (a bit like a giant cookie cutter) has <u>sharp blades</u> (for cutting) and <u>round-edged blades</u> (for creasing). These are formed into the right shape, so that they match the design of the net.

6) The die is mounted on a <u>strong plywood</u> base and <u>pressed down</u> onto the card.

7) You can cut through <u>many layers</u> of material at a time, so you can make <u>lots</u> of nets <u>very quickly</u>.

Cube net

1) A die cutter <u>presses</u> out the <u>net</u> from the sheet of material, using a <u>sharp blade</u> specially shaped to the outline of the net.

2) <u>Creases</u> can be made (along the lines where the material will be folded) by <u>rounded creasing bars</u>.

cutter   card   creasing bar

3) You have to make a <u>blade</u> especially to <u>match your net</u>, so die cutting is <u>expensive</u>, but it's great for making <u>large quantities</u> of nets with complicated designs.

8) Die cutting machines can be used for <u>loads of different things</u>, not just making nets. They can make <u>stencils</u> or cut out <u>particular shapes</u> and <u>designs</u>.

Die cutting — for when you roll a six but only need a three...

It's all about cutting on these two pages — useful to know so that you can pick the right tools for whatever project you're working on. Remember that die cutters have lots of different uses — they're not just used for nets.

Q1 Describe what the following equipment could be used for:

    a) Circle cutter   b) Perforation cutter   c) Scalpel   [3 marks]

Q2 Name an appropriate tool for cutting an intricate CAD design.   [1 mark]

Q3 What is scoring used for?   [1 mark]

Q4 Give one disadvantage of using die cutting.   [1 mark]

Printing Techniques

I can tell you're desperate to learn all about the different <u>printing methods</u>. So here are two pages absolutely <u>jammed full</u> of fun printing facts. Who knew revision could be so <u>exciting</u>...

Lithography and Offset Lithography Use 'Oily' Ink

1) Lithography uses an <u>oil-based ink</u> and <u>water</u> and works on the principle that <u>oil and water don't mix</u>.

2) <u>Ultraviolet light</u> is used to transfer the image onto a smooth <u>aluminium printing plate</u> — the <u>image area</u> gets coated with a chemical that <u>attracts</u> the <u>oily ink</u> but <u>repels water</u>.

3) So the <u>image</u> area holds <u>ink</u> and the <u>non-image</u> area holds <u>water</u>.

4) In <u>offset lithography</u>, the image is printed onto a rubber '<u>blanket</u>' cylinder which squeezes away the water and transfers the ink to the paper.

5) Lithography and offset lithography are <u>fast</u> ways of printing and they give you a <u>high-quality</u> product.

6) They're great for print runs of <u>1000 copies or more</u> — so you can print <u>books</u>, <u>newspapers</u>, <u>magazines</u>, <u>packaging</u>, etc.

Offset Lithography

cylindrical printing plate

blanket cylinder

paper

Flexography Uses a Flexible Printing Plate

Printing plate

1) <u>Flexography</u> uses a printing plate made of <u>flexible rubber or plastic</u>. The image <u>sticks out a bit</u> from the plate.

2) You can print onto <u>different surfaces</u> using flexography — they don't have to be completely flat either. This means you can print onto things that aren't totally smooth, like <u>cardboard</u>, or other packaging such as <u>plastic bottles</u>.

3) It's <u>quicker</u> than lithography and the printing plates <u>last longer</u>.

4) Flexography is used for <u>large</u> print runs (over 5000) like <u>packaging</u>, <u>wallpaper</u> and also <u>carrier bags</u>.

Gravure Uses an Etched Printing Plate

1) Gravure uses an <u>etched</u> brass printing plate — meaning the image is <u>lower</u> than the surface of the plate and the ink fills the etched bits.

2) It's <u>expensive</u> to set up but it's <u>really fast</u> and ideal for <u>very large print runs</u> (a million copies or more). The products are <u>higher quality</u> than ones printed using lithography.

3) Gravure is used to make products such as <u>postage stamps</u>, <u>photos</u> in books and catalogues, and <u>colourful magazines</u>.

Screen Printing Uses... a Screen

1) In screen printing, a <u>stencil</u> is put under a <u>fine mesh screen</u>, and <u>ink</u> is spread over the top. The ink goes through the stencil and prints onto the material below.

2) It's a <u>low-cost</u> process, ideal for <u>short</u> print runs of up to a few hundred copies where <u>fine detail</u> isn't needed.

3) You can use it to print onto <u>various surfaces</u> (e.g. paper, card, fabric) — so it's great for printing <u>posters</u>, <u>estate agents' signs</u>, etc. It can also be used on <u>textiles</u> — take a look at p.84.

Printing Techniques

Digital Printing Doesn't Use Printing Plates

1) Digital printing is done using <u>inkjet</u> and <u>laser</u> printers.

2) You don't have to make any printing plates, so it's <u>less fiddly</u> than many other methods.

3) There are <u>no set-up costs</u> apart from buying a printer and ink/ toner cartridges (which will need replacing when they run out).

4) Digital printing is <u>expensive</u> per sheet but for <u>short</u> print runs (hundreds of copies) it's <u>cheaper</u> than setting up the plates for another printing process, e.g. lithography.

5) It's used to print <u>posters</u>, <u>flyers</u>, <u>digital photos</u>, etc.

Digital Printers Use Four Colours in Layers

1) Digital printers use four colours — <u>cyan (C)</u>, <u>magenta (M)</u>, <u>yellow (Y)</u> and <u>black (K)</u>. These are also referred to as **CMYK** and are known as <u>process colours</u>.

2) Anything that's printed in colour is made up of a <u>mixture</u> of these colours.

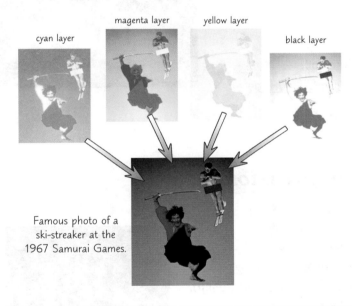

cyan layer   magenta layer   yellow layer   black layer

Famous photo of a ski-streaker at the 1967 Samurai Games.

> The 'K' stands for <u>'key'</u> — it means black. You can make black by <u>mixing</u> the other three colours, but using <u>black ink</u> usually looks better, and it works out <u>cheaper</u> if you're printing a lot of black.

3) When the computer is instructed to print, the printer recognises the required colour and adds <u>layers</u> of cyan, magenta, yellow and black to make the final colour.

4) Some printers use special <u>spot colours</u> (e.g. PANTONE® colours) as well — to print particular colours that <u>can't be achieved with CMYK</u>.

5) There's no set <u>order</u> in which the colours are put onto the paper. Most printers stick to the CMYK order while others go from <u>lightest to darkest</u> — **YMCK**.

Lithography Also Uses These Four Colours

1) Lithography (see previous page) uses <u>four</u> aluminium <u>printing plates</u> — that's one plate for each of the <u>CMYK</u> colours.

2) The image is built up in <u>layers of colour</u> (as above).

Lithography — try saying that with your mouth full of toffee...

See, wasn't that fascinating? Now you know everything you need to know about different printing techniques, and why pictures look funny when you're running low on ink — it's all down to CMYK.

Q1   Which printing method uses a plate made of rubber or plastic?   [1 mark]

Q2   Name the two printing techniques that are suitable for printing magazines.   [2 marks]

Q3   Why is digital printing used for short print runs?   [1 mark]

Q4   What colours do the letters CMYK stand for?   [2 marks]

Paper and Board Finishes

Once you've printed your product, you can choose a finish to... you've guessed it... finish your product. There are lots of different finishes to choose from, so you need to think about which properties you'll need.

Print Finishes can Make Your Product Look Top Quality

After the colours have been printed, you can use a finish on your product.

1) Finishes can improve the look of your image. Your final product will look nice and professional — so people will be more likely to buy it.
2) They help to protect your product from being damaged.
3) But adding a print finish can be expensive.

Laminating Means Sandwiching in Plastic

1) Laminating means sandwiching a document, e.g. a menu, between two layers of plastic.
2) The laminating machine heats the plastic and seals it together.
3) Laminating business cards, menus and posters makes them last longer without getting damaged.

4) It's a quick and effective way to finish a piece of work on paper or thin card.
5) Many packaging materials are a lamination of different papers, cards, plastics and aluminium foil.
6) Laminating can also apply plastic to just the printed surface — i.e. only one side of the product.
7) It can give a shiny or a matt finish.

Embossing Leaves a Raised Impression

1) Embossing means pushing a shaped die into the back of the material to leave a slightly raised impression on its surface.
2) It's often used to draw attention to a particular bit of a product, e.g. the title of a book, a logo or an image.
3) It's an expensive process but it adds texture and can suggest quality.
4) Industrial embossing machines use a rolling die to emboss the same pattern onto large amounts of paper and card.

Foil Application Makes Things Look Fancy

1) Foil application (or foil blocking) means using heat and pressure to print metal foil onto certain areas of a product.
2) Like embossing, it's used in packaging to draw attention to a logo or brand name, and to give the impression of a quality product — but it's expensive.
3) It's also used on greetings cards, book titles and wrapping paper.

Make sure embossing leaves an impression on your memory...

So now you know how products get some of those fancy finishes, but you're not done yet — there's more to learn...

Paper and Board Finishes

Varnishing Makes Things Shiny or Matt

1) Varnishing is used to make things look <u>smooth</u> and <u>glossy</u> or <u>matt</u>, so they look more exciting and high-quality.

2) You can varnish the <u>whole product</u>, e.g. <u>playing cards</u> — this makes them slide over each other.

3) Or you can varnish <u>specific areas</u> (e.g. <u>titles on book covers</u>) to draw attention to them — this is called <u>spot varnishing</u>.

4) Varnish <u>can't be written on</u>, so varnish is sometimes <u>only</u> <u>applied to one side of the paper</u> — e.g. postcards.

UV Varnishing

1) In <u>UV varnishing</u>, a varnish is <u>applied to the surface</u> of the paper. The varnish is then <u>cured under UV light</u> — this makes it feel <u>dry to the touch</u>.

2) A <u>glossy</u>, shiny finish or a <u>matt</u> finish can be applied.

3) UV varnishing can be applied to the <u>whole page</u> (<u>flooding</u>) or to <u>certain areas</u> (<u>spot UV varnishing</u>).

4) UV varnishing is a good finish for items that are going to be <u>handled frequently</u> as it protects the paper. It's often used for <u>business cards</u> and <u>magazine covers</u>.

5) The <u>varnish cures very quickly</u> when it is exposed to UV light. This <u>speeds up</u> the printing process and <u>reduces production times</u>.

1) The varnish is sprayed onto the product.

Varnish

UV lamps

UV varnishing is also known as UV coating.

Product to be finished

Conveyor belt

Direction of product movement

2) The varnish is cured under UV light.

Hope this section didn't finish you off...

Q1   Name a product that might use the following finishes and suggest why.

     a) laminating   [2 marks]

     b) foil blocking   [2 marks]

Q2   a) What is embossing?   [1 mark]

     b) Why might it be used?   [1 mark]

Q3   Why is varnish sometimes only applied to one side of paper?   [1 mark]

Q4   a) Name the varnishing process that uses UV light.   [1 mark]

     b) Suggest why using UV light is useful when producing varnished products on a large scale.   [1 mark]

Uses of Wood, Metals and Polymers

By now you should know all about paper and board. Now it's the turn of wood, metal and polymers...

Look at the Properties of a Material to See What it'll be Good at

1) As you know from Section Two, different materials have different physical properties.

2) When making a product for a specific purpose, the materials that it's made from need to have properties that are suitable for that purpose.

3) Here are some examples of products made from wood, metal and polymers and some of the reasons why these materials are suitable choices...

Have a look back at pages 14-15 for a reminder of some of the different properties materials can have.

Traditional wooden toys

1) Wooden toys are often made from hardwoods such as beech, oak and maple.

2) These woods are hard and durable, so they can withstand a lot of force. This makes them safe for children as they won't break and splinter.

3) These woods have an attractive appearance or they can be painted in bright colours.

Flat pack furniture

1) Flat pack furniture is often made from composites such as MDF (see p.23).

2) MDF is strong so it can withstand the forces that are applied to furniture.

3) MDF has a uniform texture (no knots or grain) and it can be shaped. This is ideal for flat pack furniture which needs to be cut to shape, drilled in specific locations, etc.

4) MDF can also have finishes applied to it, resulting in attractive furniture.

There are lots of other reasons why a material may be chosen for a product (see p.34-35). E.g. one of the main advantages of MDF is that it's cheap.

Cooking utensils

1) Cooking utensils can be made from stainless steel (see p.18). Stainless steel is strong and doesn't easily rust. It also has a high melting point and is safe for use with food.

2) This makes it suitable to use in the kitchen, where it will be exposed to moisture, heat and food.

Hand tools

1) Hand tools (e.g. saws, chisels, etc.) are typically made from tool steel (a steel containing up to around 1.5% carbon).

2) Tool steel is hard, tough and strong.

3) These properties make tool steel suitable for hand tools as they can cut other materials without breaking.

Polymer seating

1) Seats like these are often made from thermoforming plastics like polypropylene.

2) Polypropylene is available in lots of colours and is easily moulded into comfortable seat shapes.

3) It's quite tough, flexible and strong, so polypropylene seats can withstand people sitting down on them over and over again.

4) Polypropylene is resistant to moisture so it's suitable for outdoor use (however it can be affected by UV light — see the next page).

Electrical fittings

1) As you saw on p.19, electrical fittings are often made from urea formaldehyde.

2) This thermosetting plastic can be moulded into the required shape.

3) It'll stay hard and rigid even if it's heated again. This means it's heat and fire-resistant.

4) It's also an electrical insulator, which is essential for safety reasons.

"That's as useful as a chocolate teapot..."

...might be something your grandma says, but it's a classic example of a product made from a material which has properties that aren't suitable for its purpose. Often, you just need to use your common sense — you wouldn't make a light switch out of an electrical conductor or an umbrella out of a material that lets water through...

Uses of Wood, Metals and Polymers

The Properties of Materials can be Modified

1) Sometimes materials <u>don't</u> have the <u>exact properties</u> that you want them to. Annoying.

2) Thankfully the properties of some materials can be <u>modified</u>. This can make them suitable for <u>different purposes</u>.

3) You can modify the properties of some <u>woods</u>, <u>metals</u> and <u>polymers</u>. Here are a few examples:

Wood can be Seasoned to Strengthen it

1) As you saw on page 44, <u>seasoning wood</u> involves <u>drying</u> it. It takes place after trees have been <u>cut down</u> and had the <u>bark removed</u>.

2) <u>Reducing</u> the amount of <u>moisture</u> in the wood makes it <u>stronger</u>. It also makes it less likely to <u>rot</u> or <u>twist</u>.

3) If wood is <u>not seasoned</u> before it is used in a product then it can <u>change shape</u>. For example, if you put unseasoned wood in a <u>dry environment</u>, it will <u>dry too quickly</u>. This can cause the wood to <u>shrink</u>, which can <u>damage the wood</u> and <u>reduce its strength</u>.

4) <u>Unseasoned wood</u> is often used for <u>timber-framed buildings</u> though. This is because it is <u>easier to work</u>, it's <u>cheaper</u> and can be sourced in a <u>wider variety</u> of <u>sizes</u> than seasoned wood.

Annealing can Make Metals More Malleable

1) <u>Heat-treating</u> metals can <u>change</u> their properties.

2) For example, <u>annealing</u> involves <u>heating</u> a metal (often until it <u>glows red</u>) and leaving it to <u>cool slowly</u>.

3) This makes the metal <u>softer</u>, <u>more ductile</u> and <u>less brittle</u>.

4) Softening the metal makes it more <u>malleable</u>. This makes it easier to <u>bend</u> and <u>shape</u>.

Stabilisers can Protect Polymers from UV Light

1) <u>UV light</u> from the sun or artificial light sources can <u>damage polymers</u> by causing <u>changes</u> to their <u>chemical structure</u>.

2) Polymers exposed to UV may <u>change colour</u>, <u>fade</u>, <u>lose their strength</u> and become <u>brittle</u>. They can also get a <u>powdery residue</u> on their surface — this is known as '<u>chalking</u>'.

3) This can be <u>prevented</u> by adding a <u>chemical</u> called a <u>UV stabiliser</u> to the polymer — this <u>protects</u> the <u>structure</u> from <u>UV</u>.

4) This means the polymer can be used <u>outside</u>, e.g. for garden furniture and stadium seating.

UV stabilisers — great for learning to cycle at a rave...

Q1 Bill is making a traditional wooden toy using beech.
Give two properties of beech that make it a suitable material to make the toy from. [2 marks]

Q2 Some cooking utensils are made from stainless steel.
Give two properties of stainless steel that makes it suitable for use in cooking utensils. [2 marks]

Q3 Flat pack furniture is often made from MDF.
State two properties of MDF that make it a good material for flat pack furniture. [2 marks]

Q4 What property of wood can be modified by seasoning it? [1 mark]

Q5 a) What is the name of the process used to soften metal? [1 mark]

  b) Outline what happens during this process. [2 marks]

Stock Forms and Standard Components

Timber, metal and polymers can be bought in different shapes (known as stock forms) and sizes.

Timber is Available in Planks, Strips, Mouldings and Boards

Timber can be bought in lots of forms:

1) Timber comes in different sizes — for example planks and strips are sold in standard thicknesses, widths and lengths, e.g. 47 x 100 x 2400 mm.

2) Planed Square Edge (PSE) timber has its rough surfaces shaved off with an electric planer to give smooth surfaces and sharp corners.

3) Rough sawn timber is not smoothed after it's cut — so it's cheaper and useful for construction work where it won't be visible.

4) Mouldings are strips that come in a range of cross-sections. They're used for skirting boards, door frames, picture frames, etc.

5) Manufactured boards (MDF, plywood, blockboard, etc.) are usually available in 2440 x 1220 mm sheets (although you can get smaller sizes). Common thicknesses are 4, 6, 9, 12, 15 and 18 mm.

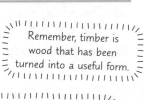
Remember, timber is wood that has been turned into a useful form.

Mouldings with circular cross-sections are available in different diameters.

diameter

Metals Come in Loads of Shapes and Sizes too...

1) Metals are commonly available in a wide range of shapes and sizes, because it can be very difficult to convert one shape to another.

2) This means that the manufacturers can buy roughly the right shape to start working with.

For example, the shafts of these golf clubs were made by using tubes of metal.

sheet
strip
bars
rod
tube

angle
U-shaped channel
I-shaped girder

Like timber, metals are available in a range of thicknesses, widths and lengths. Lengths of metal with a circular cross-section can be bought in different diameters.

...and so do Polymers

Here are some of the different forms that polymers come in...

You buy granulated and powdered polymers by weight.

Sheets, tubes and rods can be cut to size and bent (see p.69). Sheets can also be used in drape forming (see p.71).

Foam is used for protective packaging and making models and mock-ups.

Films are good for vacuum forming (p.70), and for use as windows in packaging.

Granules can be melted down and used in casting and moulding (see p.69-70).

Powders can be used in some types of moulding, as coatings (see p.73) or in 3D printing (see p.68).

Some polymers (e.g. foam and rods) are available in standard widths, lengths and thicknesses (or lengths and diameters). Some thin polymers (e.g. sheets and films) are available in different gauges (standard thicknesses) — gauge gives an indication of how strong the polymer will be.

I don't know about you, but I feel spoilt for choice...

You can buy materials in loads of different forms. Now onto some bits and bobs that you can use with them...

Stock Forms and Standard Components

You need to know about some <u>pre-manufactured standard components</u> (see p.50) and how they can be used.

Screws and Bolts are Temporary Fastenings

1) <u>Screws</u> and <u>nuts and bolts</u> are <u>temporary</u> ways of joining things together — you can take them <u>apart</u> if you need to.

2) They are usually made from steel, brass or stainless steel, and are '<u>self-finished</u>', plated with <u>zinc</u>, <u>brass</u> or <u>chrome</u>, or coated in a <u>black varnish</u>.

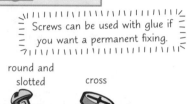
Temporary fixings are designed to be taken apart, and put back together if necessary. Permanent fixings are... permanent.

3) It's important to use the <u>right sort</u> to make sure that the job <u>looks good</u> and <u>holds together properly</u>.

Screws

1) Screw threads (the twisty bit around the outside of the screw) <u>grip</u> the material, making a <u>strong</u>, <u>tight fixing</u> in wood, metal and plastic.

2) They are used for lots of jobs, e.g. for access to the batteries in <u>children's toys</u>, <u>self-assembly furniture</u>, fixing <u>shelves</u> and <u>mirrors</u> to walls, and fixing <u>hinges</u> to doors.

There are <u>different types</u> of screws for use with <u>wood</u>, <u>metals</u> and <u>plastics</u>:

Screws can be used with glue if you want a permanent fixing.

1) <u>WOODSCREWS</u> often require '<u>pilot</u>' holes to be drilled before the screw is inserted.

2) As the screw is turned by a <u>screwdriver</u>, the thread pulls it into the wood.

3) Different types of <u>head</u> are available for different jobs, e.g. <u>round</u>, <u>countersunk</u>, <u>slotted</u> and <u>cross heads</u>.

countersunk and slotted   round and slotted   cross

1) <u>SELF-DRILLING SCREWS</u> have <u>hardened threads</u> and a <u>tip</u> like a <u>drill bit</u>.

2) They are designed to <u>drill their own threaded holes</u> in hard materials such as <u>metals</u> and <u>hard plastics</u>. They don't require a <u>pilot hole</u> to be drilled before they're inserted.

1) <u>MACHINE SCREWS</u> have a straight (not tapered) <u>threaded shank</u> and are used with <u>washers</u> and <u>nuts</u> or in a hole with <u>matching threads</u>.

2) Their heads vary (round, countersunk, etc.). Some are tightened with a <u>screwdriver</u> (cross and slotted types) and some with an <u>Allen key</u> (socket head).

3) They are often used to join <u>metal</u> parts together.

Straight shank

Nuts and Bolts

1) Nuts and bolts are used to join <u>thin materials</u> like sheets of <u>metal</u> or <u>plastic</u>.

2) They're only useful if you can get to <u>both sides</u> of the materials you're joining.

3) They're used to join parts of <u>machinery</u>, e.g. in cars and bridges, and to hold <u>moving parts</u>, e.g. in play swings.

4) <u>Bolts</u> are similar to machine screws (see above) but usually have a <u>square</u> or <u>hexagonal</u> head that can be tightened with a <u>spanner</u>.

bolt

nut

head

shank

thread

Try not to screw it up in the exam...

Q1   Name a material that screws are often made from.   [1 mark]

Q2   What are machine screws commonly used for?   [1 mark]

More Standard Components

More components for you now. First up, rivets and hinges, then onto knock-down fittings...

Rivets are Mainly Used for Joining Sheet Metal

1) A rivet is a metal peg with a head on one end.
   Rivets are mostly used for joining pieces of metal.

2) A hole is drilled through both pieces of metal and the rivet is inserted with
   a set (hammer-like tool). The head is held against the metal whilst the
   other end is flattened and shaped into another head with a hammer.

3) 'Pop' (or 'blind') rivets are now very common. They can be used where there is only access to
   one side of the material (hence 'blind' rivet). It's a fast and easy method of joining sheet metal.

standard rivets

How pop rivets work

1. The metal pin is inserted through the hole in the centre of the pop rivet.

2. Both rivet and pin are placed in the hole through the materials.

Rivets are permanent fixings.

3. The pin is pulled tight with a riveter (or pop gun) till it snaps off.

4. Pulling the pin tight squashes and spreads the end of the rivet, forming a head on the 'blind' side.

There are Four Main Types of Hinge

Hinges are available in steel, brass and nylon (a polymer), and can be coated to match a piece of furniture.
The part of the hinge that moves is called the knuckle.

Butt hinges

1) Butt hinges are the most common type of hinge used for doors.

2) One part of the hinge is set into the door and the other part is set into the frame.

3) They're available in brass or steel.

Tee hinges

1) Tee hinges are often used outside for things like shed doors or garden gates. The longer 'strap' allows the hinge to support a greater weight.

2) They're often covered in black enamel.

Pivot hinges

1) Pivot hinges allow you to lift a door off its frame.

2) The hinge is made from two parts that fit together. One part is screwed to the door and the other is screwed to the door frame.

Flush hinges

1) Flush hinges are screwed directly onto the surface of the wood, so they're easier to fit than butt hinges.

2) They're usually used for lightweight jobs.

I'm beginning to feel unhinged...

Have a quick look around the room you're in and see if you can identify any hinges that you see.
Now go to the next room and do the same — then the next... It'll keep you entertained for hours...

More Standard Components

Knock-Down (KD) Fittings are Temporary Joints

1) Knock-down fittings are <u>temporary</u> fittings that allow furniture to be <u>assembled</u> and <u>taken apart easily</u>. They're usually used for cheap '<u>flat-pack</u>' furniture.
2) They're very <u>fast</u> to use, but are nowhere near as <u>strong</u> as glued joints.
3) Most types are assembled with <u>screwdrivers</u> or <u>Allen keys</u>.
4) There are different types for <u>different joints</u>...

Scan fittings

1) A <u>cylinder</u> with a <u>screw thread inside</u> is put into a ready-made <u>hole</u> in a part.
2) A <u>screw</u> is then put into <u>a second part</u> and screwed <u>into the cylinder</u>, pulling both parts <u>together</u>.

They are sometimes called barrel nut and bolts.

3) It's used with materials like <u>plywood</u> where the screw wouldn't <u>grip</u> well (e.g. in tables to attach the legs to the frame).

CAM locks

1) A <u>disk</u> with a <u>slot</u> in fits into one part. A <u>peg</u> is screwed into a second.
2) When the parts are <u>pushed together</u>, the peg is pushed into the slot in the side of the disk. <u>Turning</u> the disk <u>grabs</u> the <u>peg</u> and pulls the parts <u>tightly together</u>.
3) It's used to join the <u>sides</u> of cabinets to the <u>top</u>. It looks quite <u>neat</u> (it's mostly hidden <u>inside</u> the parts).

Single and two-piece blocks

1) These are <u>plastic blocks</u> used to join <u>parts</u> at <u>right angles</u>.
2) They have holes in them to fit <u>screws</u> through, and screw into ready-made <u>holes</u> in the parts being joined.

Single and two-piece blocks are sometimes called modesty blocks.

3) <u>Two-piece blocks</u> are first screwed into the parts, and then joined together with a <u>bolt</u>, making the joint straight and strong.
4) They can be easier to attach in <u>tight corners</u> than single piece blocks, and can be <u>undone</u> if required.
5) They're often found holding up <u>shelves</u> in wardrobes.

Dowels

1) Dowels are <u>wooden rods</u> that are pre-cut to length.
2) They're used in <u>dowel joints</u>. Holes are drilled into the <u>ends</u> and <u>sides</u> of <u>parts</u>, and the dowels fit into them like a <u>peg</u>.
3) They're usually used with <u>glue</u>, and give more <u>contact area</u> to the joint to make it stronger. The dowels have little <u>channels</u> in them to make space for the glue. Gluing makes them a <u>permanent fixing</u>.
4) They're often used to attach <u>shelves</u> to cabinets.

Standard joke components — an Englishman, an Irishman...

Q1   Jack is making a door for a small kitchen cupboard. Suggest what type of hinge he could use.   [1 mark]

Q2   Give one advantage and one disadvantage of using knock-down fittings.   [2 marks]

Shaping Materials — Hand Tools

Wood, metals and polymers can be cut, shaped and smoothed. This can be done using hand tools. These are (unsurprisingly) tools that work by hand (i.e. not machine tools or power tools).

Saws are the Main Cutting Tools

1) There are different saws for different materials. Saws with lots of little teeth close together (fine pitch) are usually for hard materials and saws with bigger teeth further apart (coarse pitch) are for soft materials. Using the wrong saw could damage the material (or hurt you — it could 'jump' or catch unexpectedly).

Rip saw — for cutting wood along the grain

Tenon saw — for making straight cuts in small pieces of wood

Hacksaw — for metals and plastics

Coping saw — for cutting curves in wood or plastic

2) Tips for using a saw:
- Try to use the whole length of the blade so that some teeth don't wear out faster than others.
- Don't press too hard or the saw might jam in the material. This will damage the blade, the material or even you when you try to pull it out.
- Rough edges from sawing can be tidied up by sanding or planing.

Sounds obvious, but saws need to be kept sharp — either by sharpening or replacing the blade.

Chisels are Used for Shaping

Chisels are used to cut away and shape wood and metal.

1) Wood chisels come in different profiles for making different shapes. You hit them with a mallet.

2) Gougers are used for sculpting.

3) For metal, you need cold chisels. These are hit with a hammer.

Planes, Files and Abrasive Papers can Shape and Smooth

1) A bench plane has an angled blade that shaves off thin layers of material. It's used on wood for removing material (shaping).

hand file
triangular file
half round file
flat file

2) Files have hundreds of small teeth to cut away at a material. Different 'cuts' make them suitable for different processes: rough cuts are for removal of material, fine cuts are for finishing (final smoothing). They come in different profiles — to make different shapes. Most files are meant for metals and wood. There are special ones with very coarse teeth called cabinet rasps that can be used on wood.

3) Abrasive papers such as sandpaper (or glass paper) are rubbed against the material you want to shape or smooth. They are often wrapped around a sanding block so an even pressure can be applied and to give you better grip. Abrasive papers are available in a range of grit sizes — from coarse to fine. Different types are available for use on wood, metal, and polymers.

Wet and dry (silicon carbide) paper is a type of abrasive paper that is often used on plastics and metals. It can be used wet — this stops it getting full of dust.

My file slipped — now my finger is saw...

It's really important to select the right tool for the job and the material that you're using. If you don't, it's likely that you'll end up not using the tool properly to try and get it to do what you want — and that can be dangerous.

Shaping Materials — Hand Tools

Drills Make Holes (no kidding...)

1) A <u>bradawl</u> can be used to help you drill in the right place. You <u>press</u> it into the material to make a <u>little dent</u> where you're going to drill — this <u>stops</u> the <u>drill bit</u> from <u>slipping</u>. Bradawls can only be used on <u>wood</u> and <u>plastic</u>. A <u>centre punch</u> is used for <u>metal</u> — you <u>hit</u> it with a <u>hammer</u> to make a dent in the metal's surface.

bradawl

centre punch

2) Depending on how hard the material is, you can do the actual drilling with a <u>brace</u>, a <u>hand drill</u> or a <u>power drill</u>.

3) All drills work by rotating a <u>drill bit</u> against the material.

brace

hand drill

<u>Twist bits</u> are used to drill <u>small holes</u> in wood, metal or plastic. <u>High speed steel</u> (HSS) twist bits are used on metals and plastics.

<u>Flat bits</u> are used on <u>wood</u> and <u>plastics</u> to drill <u>large</u>, <u>flat-bottomed</u> holes. <u>Forstner bits</u> make similar holes, but can only really be used on <u>machine</u> drills (see p.67).

<u>Countersink bits</u> make holes for <u>countersunk screw heads</u> to sit in.

<u>Hole saws</u> are a bit like <u>round saws</u> that are used in a drill. They're used to make <u>big holes</u> in <u>thin material</u>.

4) Drill bits can get really <u>hot</u>, so be <u>careful</u> when taking them out of the material.

Hand Tools Need to be Used Safely

When you're using hand tools:

- Make sure the object you're working on <u>can't move</u> — hold it firmly in a <u>bench hook</u>, <u>vice</u> or <u>clamp</u>.
- Keep your <u>fingers</u> out of the way.
- Be aware of any <u>exposed blades</u>, and always carry tools <u>safely</u>.
- It's a good idea to wear <u>goggles</u> to stop <u>shavings</u> getting in your <u>eyes</u>.

Get every bit of this learnt — you know the drill...

Q1   Which of the following types of saw is best for cutting metals?
       a) rip saw   b) tenon saw   c) hacksaw   d) coping saw   [1 mark]

Q2   Rob wants to remove a thin layer of wood from the bottom of a door he is fitting.
       Suggest a hand tool that he could use to do this.   [1 mark]

Q3   The photo on the left shows two drill bits.

       a) Name drill bit A and B.   [2 marks]

       b) What is each drill bit used for?   [2 marks]

       c) What tool could be used to make a pilot hole in a piece of wood that is going to be drilled?   [1 mark]

Q4   Give two pieces of safety advice you would give to someone using a rip saw for the first time.   [2 marks]

Shaping Materials — Power and Machine Tools

Finally — the bit you've been waiting for — using power tools to cut and shape stuff. Ooooo.
It doesn't get much better than this. Before you get stuck in though, a little safety announcement...

Safety is Really Important

There's more about safety on page 118.

Power tools are hand-held motorised tools. You need to use them safely...

1) Before using power tools, do a visual check for any loose connections and run your hand
   along the lead to check for any cuts in the insulation (when it's not plugged in, of course).
   Check that the blade or drill bit or whatever is attached correctly and tightly.

2) You can use an RCD (Residual Current Device) to help prevent electric shocks. The power
   tool plugs into the RCD, which you plug into the socket. If you accidentally cut through
   the lead of the power tool, the RCD cuts off the electricity supply straight away.

3) Wear a mask or fit an extraction hose if the tool's going to produce a lot of dust.
   Always wear safety glasses and make sure clothing can't get caught.

4) Clamp your work down firmly so it can't slip or move.

5) Make sure you know where the stop buttons are before you start.

6) When you've finished, make sure the tool has stopped moving before you put it down.

Power Tools can be Used to Cut, Shape and Smooth Materials

Like with hand tools, there's a huge range of power tools for shaping materials. Here are a few examples:

Routers

1) Hand-held routers have a spinning cutting tool that cuts away wood.
   They're used to make features like slots, grooves and fancy edges.

2) A router is usually used with a fence — a thing that guides
   the router and keeps it in the right position (see next page).

3) You can get different cutting tools to make different shapes.

You met CNC routers on page 5.

Planers

1) A planer is used like a bench plane to remove shavings of wood —
   either to reduce the material to the required size, or for rough shaping.

2) The advantage of a power planer is that it takes much less effort
   and is much faster — but it's not as accurate as a bench plane.

3) Before you start, check the blades are sharp and replace them if not.

4) Don't start with the cutting tool on the wood — rest the flat front base on the wood and
   start to push it forward when it reaches full speed.

5) Make sure you're well-balanced and use two hands to hold it — one on the front handle
   and one on the trigger switch.

Jigsaws

1) A jigsaw has interchangeable blades and variable speeds.

2) You can make straight or curved cuts in all materials, but it's quite slow.
   A fence helps you make straight cuts.

3) Make sure the blade is secured tightly and the correct type of blade is installed
   for the material. The teeth should face the front of the saw, and you should
   push it forwards (away from you). Don't start cutting until the blade is at full speed.

Sanders

1) Electric power sanders can be used to smooth wood.

2) They work by moving abrasive paper at high speeds.
   This is a lot quicker and easier than sanding by hand.

3) Different types of power sanders work in different ways. For example,
   belt sanders have a loop of abrasive paper which rotates at high speeds.

4) Most sanders have a dust bag or an extraction hose to remove the dust that they produce.

Shaping Materials — Power and Machine Tools

Machine Tools are Great for Bigger Jobs

1) <u>Machine tools</u> generally do the <u>same jobs</u> as power tools — but they're often more <u>accurate</u> and better suited to <u>bigger tasks</u>.

2) Machine tools are usually <u>stationary</u>, often <u>bolted</u> to the workbench or the floor. They're usually attached to a <u>dust extractor</u> too for when <u>wood</u> is being <u>machined</u>. These tools can be used to process large quantities of material <u>accurately</u> and <u>quickly</u>.

Machine Tools can be used to Cut, Drill and Sand

Sanding Disc

1) A <u>sanding disc</u> spins a <u>disc of abrasive paper</u> which the material's pushed against.

2) It's used for accurately <u>removing material</u> to a line.

3) Different types of abrasive paper can be used on <u>wood</u>, <u>metal</u>, and <u>plastics</u>.

Saw Bench

The <u>saw bench</u> has a circular blade and is used to cut <u>wood</u> and boards like <u>plywood</u> to size. It makes <u>straight cuts only</u>.

Saw bench

There are more machine tools on the next page.

Band Saw

The <u>band saw</u> has a blade in a <u>long flexible loop</u> and is normally used to cut <u>wood</u>, but <u>special blades</u> can be bought for use on <u>plastics</u> and <u>softer metals</u>. The blades come in different <u>widths</u> and can be used for <u>straight</u> or <u>curved</u> cuts.

Pillar Drill

A <u>pillar drill</u> (or <u>pedestal drill</u>) is used to make round holes. They can be used on <u>all kinds</u> of materials, depending on the drill bit used (see page 65).

Fences are Attached to Tools to Improve Accuracy

1) Fences are attached to <u>power</u> and <u>machine tools</u> to help <u>guide</u> either the tool or the material.

2) In the example on the right, the <u>fence</u> runs along the <u>edge</u> of the material, keeping the router the <u>same distance</u> from the edge <u>at all times</u>. This keeps the cut <u>straight</u>.

3) Fences can be set up to <u>repeat</u> the same action over and over again — ensuring <u>consistency</u> between the products in a batch (see p.38).

Router

Fence attached to router

The band saw — the most musical of the machine tools...

When using tools to cut a material, you still need to check the part you've made is within tolerance (see p.40).

Q1 Susan is making a door. Suggest the power tool that she should use to:

a) cut the door down to the right height. [1 mark]

b) make a groove 10 cm in from the edge of the door. [1 mark]

Shaping Techniques

There are loads of <u>techniques</u> for getting materials in the <u>right shape</u>. The best one for the job depends on the <u>shape</u> you're <u>aiming for</u> and the <u>type of material</u> you're shaping. Here are some to get you started...

Milling Machines Remove Thin Layers of Material

1) <u>Milling machines</u> remove material one <u>thin layer at a time</u> to produce the required size or shape.

2) These machines can be set to a particular <u>cutting speed</u> and <u>depth</u>.

3) They can also be used to make a surface <u>absolutely flat</u> and can produce a very <u>accurate finish</u>.

Lathes Are For Turning

Turning means shaping a material while it rotates.

1) <u>Lathes</u> come in two types — <u>wood lathes</u> and <u>engineers' lathes</u> (also known as a centre lathe, for working metal). They're used to 'turn' materials, to make objects like these.

2) A material is <u>held</u> and <u>rotated</u> by the lathe, while a <u>tool</u> or <u>bit</u> is pressed onto the material to cut it.

3) Make sure the material is held <u>tightly</u> and <u>straight</u>. If you're turning a <u>long</u> piece of wood or metal, it needs to be held securely at <u>both ends</u> to stop it <u>wobbling</u>.

> Both milling machines and lathes can be used in <u>CAM</u>. They are <u>subtractive</u> CAM processes — material is <u>removed</u> from a larger piece of material to make the product.

Take a look at page 4 for how milling machines are used with CAM.

3D Printing is an Additive CAM Process

1) <u>Additive</u> CAM processes involve <u>adding material</u> to build up the product rather than by removing it.

2) <u>3D Printing</u> is <u>additive</u>. It works by printing <u>layers</u> of molten plastic, powder or wax until the full 3D shape has been formed.

3D printing is often used to make prototypes (p.5) and models (p.106).

3) You can <u>design</u> your ideas on screen using various <u>software</u> packages (this is an example of <u>CAD</u> — see p.4), then use a <u>3D printer</u> to convert your design into a <u>3D model</u>.

Metal Sheets can be Pressed into Shape

1) <u>Press forming</u> involves pressing metal sheets between <u>two moulds</u> with a <u>large amount</u> of <u>force</u>.

2) The <u>metal sheets</u> to be pressed are often <u>annealed</u>, as this means they're more <u>malleable</u> (see page 14).

Here's how it's done:

metal sheet — punch — die

HUGE FORCE

Press forming can be done at room temperature — the metal doesn't need to be heated to shape it.

shaped metal

1) A <u>sheet</u> of <u>metal</u> is placed between two moulds — a 'punch' and a 'die'.

2) A <u>hydraulic press</u> moves the punch <u>down</u> and a <u>large force</u> is <u>applied</u> to the metal sheet.

3) This pushes the sheet of metal into the die, producing the <u>required shape</u>. The metal sheet has been <u>pressed</u>.

3) <u>Similar</u> machines such as <u>punches</u> and <u>die cutters</u> can <u>cut out</u> parts from a sheet of material (e.g. a metal or plastic). The machine applies <u>shear forces</u> (see p.36) to deform the material in this way.

Casting is Used to Shape Materials

pattern

mould

cavity

moulding sand

1) The process of casting involves <u>molten material</u> being poured into a <u>hollow mould</u>. The material is then left to <u>cool and solidify</u> before being <u>separated</u> from the mould.

2) Before casting a mould must be made. A <u>pattern</u> (an exact replica of the item to be cast) is used to create a <u>cavity</u> in a box filled with moulding sand (special sand that doesn't stick to the material being cast). This cavity will be <u>filled with material</u> during casting.

3) <u>Moulds</u> are often made in <u>two halves</u> so they can be split open to release the cast material.

Shaping Techniques

Die Casting Can be Done on Metals and Plastics

1) Die casting is used to mould <u>metals</u> and <u>thermoforming plastics</u>.
2) The material is <u>melted</u> and poured into a <u>mould</u> (the 'die') which is in the shape of the product.
3) It is then allowed to <u>cool</u>. Once it has <u>solidified</u> it can be removed from the mould.
4) Some plastic resins can be <u>cold-poured</u> into moulds (without heating). They <u>harden</u> or <u>set</u> through a <u>chemical reaction</u>.
5) When the product is <u>removed</u> from the mould it may need to be <u>trimmed</u> to remove any <u>excess material</u>.

There are a Few Ways That Materials can be Bent

1) You can use a <u>SHEET METAL FOLDER</u> to shape <u>sheet metals</u> such as aluminium and tin.
2) The outline of the product, e.g. a box, is marked out and cut from a <u>flat</u> sheet of metal.
3) You <u>feed the metal in</u> flat, make one fold then move the material through for the next fold.
4) Corners can then be <u>joined</u> using rivets (see page 62), or by soldering, brazing, etc. (see page 71).

1) <u>Wood</u> can be bent by <u>LAMINATING</u> it. There's more on lamination on p.37.
2) <u>Thin strips</u> of wood are <u>glued together</u> and held in a <u>jig</u>, which keeps them <u>bent</u> into the <u>right shape</u>. Once the glue has <u>dried</u>, the strips of wood can be taken out of the jig, and <u>stay bent</u>.
3) Things that could be made this way include <u>rocking chair runners</u>, <u>chair</u> or <u>table legs</u> and <u>roof beams</u>.

strips of wood / jig

1) <u>LINE BENDING</u> is ideal for use with <u>acrylic sheets</u>, e.g. for making picture frames, pencil holders, etc.
2) It can be done manually or with a <u>line bender</u> or <u>strip heater</u>.
3) You rest the sheet on two bars and the element between them <u>heats</u> the plastic. You just need to position the sheet carefully, so that the line you want to bend along is <u>directly above</u> the element.
4) Once the plastic is <u>soft</u> it can be <u>bent</u>. When it <u>cools</u> it will stay in its <u>new shape</u>.

heating element

Most Metals Need to be Heated Before Bending

1) Some <u>thin</u> pieces of metal can be bent cold on a <u>jig</u> or <u>former</u>.
2) <u>Thicker</u> or harder metals have to be heated or <u>annealed</u> first (see page 59) and allowed to cool.
3) This makes them soft enough to bend easily, but the annealing process might have to be repeated as bending makes them go <u>hard</u> again — this is known as '<u>work hardening</u>'.

bending jig / BEND

All this talk of getting into shape — it's time for some exercise...

If you're asked to <u>outline</u> how to do something, make sure you include enough detail. For example, don't just say, 'using a sheet metal folder' — describe the <u>different steps</u> of the process involved to get all the available marks.

Q1 Describe what a milling machine is used for. [2 marks]
Q2 a) Why is press forming often carried out on metals that have been annealed? [1 mark]
   b) Outline the process of press forming. [3 marks]
Q3 Name a process that could be used to fold a piece of acrylic. [1 mark]

Moulding and Joining

There are plenty of ways to <u>mould</u> materials and <u>join</u> them <u>together</u> — especially <u>plastics</u> and <u>metals</u>...

Air is Sucked Out In Vacuum Forming

1) A mould is put onto the <u>vacuum bed</u>.

2) <u>Thermoforming plastic</u> sheets or films (e.g. polypropylene or HIPS) are <u>clamped above</u> the <u>vacuum bed</u> and <u>heated</u> until they go <u>soft</u>.

3) The vacuum bed is <u>lifted close</u> to the heated plastic.

4) The air is <u>sucked</u> out from under the plastic, creating a <u>vacuum</u>. The air pressure from outside the mould then <u>forces the plastic onto the mould</u>.

5) The moulded plastic is <u>cooled</u> and the vacuum bed <u>lowered</u>. The <u>cold</u> plastic is <u>rigid</u> so holds its <u>new shape</u>.

heated thermoforming plastic

vacuum bed moves up

air sucked out

mould (often made from MDF) on a vacuum bed

vacuum bed moves down

Blow Moulding...Well...Blows Air In

1) A tube of <u>softened plastic</u> is inserted into a <u>solid mould</u>.

2) <u>Air</u> is then injected which forces the plastic to <u>expand</u> to the <u>shape</u> of the <u>mould</u>.

3) This method is often used to produce <u>bottles</u> and <u>containers</u>.

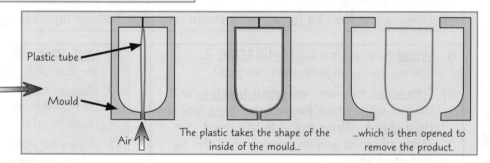

Plastic tube

Mould

Air

The plastic takes the shape of the inside of the mould...

...which is then opened to remove the product.

Injection Moulding Uses Pressure to Mould Plastics

1) Injection moulding is similar to casting (see p.68), but the molten material is forced into a <u>closed mould</u> under <u>pressure</u>.

2) The moulds are often made from <u>tool steel</u> — it's very hard but it's also quite expensive. The plastic is often melted using <u>built-in heaters</u>.

3) It can be used to make things like <u>plastic buckets</u> and <u>watering cans</u>.

Hopper

Plastic granules

Mould (or die)

Heater

Pressure

plastic melts in chamber

In both vacuum forming and injection moulding the moulds must have rounded corners and be slightly tapered (sloped) at the sides — so that the finished product can be released from the mould.

Plastics expert Brian had doubts about his new job.

Extrusion Produces Long, Continuous Strips

1) Extrusion is similar to injection moulding. It's used for <u>thermoforming plastics</u> and some <u>metals</u>.

2) The material is <u>melted</u> and forced under <u>pressure</u> through a <u>die</u>.

3) It produces long, <u>continuous</u> strips of the moulding exactly the same shape as the exit hole. It's used for products like <u>plastic-covered wire</u>, and <u>plastic and aluminium window frames</u>.

Plastic granules

Heater

The die can produce strips with different cross-sectional designs, e.g.

Pressure

Mould (or die)

Moulding and Joining

Drape Forming Moulds Softened Thermoforming Plastics

heated thermoforming plastic (soft)

mould of desired shape | placed on mould and left to cool | final product

1) A sheet of thermoforming plastic is heated until it softens.
2) The softened sheet takes the shape of the mould that it is placed on. (It's a bit like putting a table cloth on a table — it takes the shape of the table.)
3) Once the plastic has cooled it will stay that shape.

Materials Can be Bonded Together with Glue

There are all kinds of adhesives (glues) — which one you use depends on the material you're joining. Here are a few examples:

1) Polyvinyl acetate (PVA) glue is used with wood, paper and card. It dries slowly.
2) Glue guns use melted plastic to join woods and fabrics. It's especially good for modelling.
3) Solvent cement is used to join some plastics. The joint needs to be clamped whilst it sets.
4) Epoxy resin and superglue can be used on most materials. They set quickly but aren't cheap.

Chad fell for the old epoxy resin/hair gel switch-e-roo.

Soldering, Brazing and Welding are for Joining Metal

1) SOLDERING is a relatively low temperature process. It involves melting solder (made from tin and other metals) onto the components to be joined — this sticks them together when it cools and solidifies.
2) A soldering iron or a blow torch can be used for this process.

There's more about soldering on page 91.

1) BRAZING is a higher temperature process that often uses brass spelter as the joining material. It's much stronger than soldering.
2) Either a gas brazing torch, a blow torch, or a brazing attachment for an electric-arc welder is used to heat the joint.

1) WELDING is by far the strongest method of joining metal.
2) It uses a very high temperature from a gas (oxyacetylene) torch, or an electric-arc welder to melt the edges of the joint so that they flow together. A welding rod is melted at the same time to fill the joint.
3) Alternatively, metal sheets can be spot welded — this is where the metal sheets are melted and joined at a number of separate spots along their edge. This type of weld doesn't require welding rods, and is often carried out by robots on production lines (e.g. in car manufacturing).
4) Welding can also be used to fill in thinned metal or slight gaps.

Welding masks are worn to protect your eyes from the bright light and UV radiation from the welding arc. It also protects your face from heat and sparks.

welding mask

welding rod

Polymers can be Welded Too

Pressure is applied to the joined parts during cooling.

1) Polymers can also be welded together using high temperatures. Like with metal welding, the process melts the edges of the joint so that they flow together. Welding rods made from thermoforming plastic can be used to melt plastic into the join.
2) The heat required can be provided by different tools such as hot gas welding guns, lasers, etc. Heat can be generated in other ways too, e.g. a polymer can be vibrated at high speeds against the polymer it is to be joined to. The friction this produces generates heat, which welds the parts together.
3) Solvents can also be used to weld polymers. They chemically 'melt' the polymers to join them together.

You're almost at the end of Section Five — keep soldering on...

Q1 Describe the process of vacuum forming. [5 marks]
Q2 What is solder made from? [1 mark]

Treatments and Finishes

There are loads of different <u>treatments</u> and <u>finishes</u> that you can apply to <u>timber</u>, <u>metals</u> and <u>polymers</u>.

Prepare Wood Before Applying Finishes

1) Before applying <u>any finish</u>, timber needs to be <u>sanded</u> to a smooth finish using <u>abrasive paper</u> (see page 64). Sand in the direction of the grain.

2) It's often a good idea to apply a <u>sanding sealer</u> before sanding timber. Otherwise, the first coat of <u>varnish</u> or <u>wax</u> can cause loose wood <u>fibres</u> to <u>stick up</u>, and the timber will have to be <u>sanded again</u>.

3) Before <u>painting</u>, <u>bare timber</u> needs to be <u>primed</u>. Priming does three useful things:
   - <u>Fills</u> the <u>grain</u> of the timber to give the paint on top a <u>smooth finish</u>.
   - Helps <u>seal</u> the timber, so paint won't <u>soak in</u> and need to be re-applied.
   - Helps the paint to <u>stick better</u> to the timber.

There Are Many Types of Paint

Paints are used to <u>protect timber</u>, when you want to <u>cover up the grain</u> and change its <u>colour</u>. Between <u>each layer</u> of paint you <u>sand lightly</u> in the direction of the grain.

Undercoat | <u>Undercoat</u> is the first coat of paint (on top of a primer if you've used one, or on old finishes if you're reusing timber). It <u>covers up</u> any previous <u>colours</u> more <u>cheaply</u> than applying extra layers of the final 'topcoat', which is usually a more expensive paint. It helps later layers of <u>paint</u> to <u>stick</u>.

Top coats |
1) <u>Gloss paints</u> are <u>hard-wearing</u> and <u>waterproof</u> and come in lots of <u>colours</u>. They're shiny and used for things like interior woodwork. You apply them with a <u>brush</u> or <u>roller</u>, painting in the <u>direction of the grain</u>.
2) <u>Polyurethane paints</u> are even <u>tougher</u> than gloss paints. They're used for things like <u>children's toys</u>. They're often <u>sprayed</u> on for a <u>smooth finish</u>.

So, deep breath: sand, prime, sand, undercoat, sand, paint, sand, paint.

Varnish Lets You See the Grain

Many woods (especially hardwoods) have an <u>attractive grain</u> — so you might want to use <u>clear varnish</u> rather than paint as a surface finish. For example, oak furniture is often varnished.

1) Varnish can be <u>coloured</u> or <u>clear</u>, and either <u>gloss</u> (shiny), <u>matt</u> (dull) or <u>satin</u> (in between).

2) <u>Yacht varnish</u> seals the wood and makes it <u>waterproof</u>. It's pretty <u>flexible</u>, so it <u>doesn't crack</u> if the wood <u>moves</u>. So it's good for outdoor uses, e.g. doors and window frames (and yachts).

3) <u>Polyurethane varnish</u> is best for <u>interior</u> uses, e.g. stairs and skirting boards. It's very <u>hard-wearing</u>.

4) For best results, you need to apply <u>two or three coats</u>, <u>sanding</u> lightly <u>between</u> each coat.

Wood can be Treated in a Process Called Tanalising

Woods that will be used <u>outdoors</u> are often treated with a <u>wood preservative</u>. This helps to prevent <u>insect attacks</u> and <u>decay</u> of the wood, meaning it will <u>last longer</u>. Here's how it's done:

Putting the wood in a vacuum removes air from the cells of the wood — this means the preservative can go there instead.

1) Seasoned timber is placed into a <u>treatment tank</u>.
2) The <u>air</u> is <u>removed</u> from the tank, creating a <u>vacuum</u>.
3) The tank is <u>flooded</u> with the <u>preservative</u>.
4) <u>Pressure</u> is applied which <u>forces</u> the preservative <u>deep</u> into wood.
5) <u>Excess preservative</u> is <u>removed</u> and the timber is left to <u>dry</u>.

Tanalised® wood is used for things like <u>outdoor playgrounds</u>, <u>fences</u>, <u>telegraph poles</u>, etc.

Treatments and Finishes

Metals Need To be Prepared Before Finishing

1) Metals have to be smoothed first. You can do this by filing. Next, you rub the metal with gradually finer grades of abrasive paper and wet and dry paper.

2) Then you need to remove grease from the surface (by rubbing with a cloth soaked in paraffin or degreaser) — to make sure the finishes will stick properly.

Metals Can be Finished for Protection and Looks

1) Some metals don't need a finish, but some can corrode without suitable protection.

2) Corrosion is where metals react with substances, become oxidised and are gradually destroyed.

3) Finishes can also improve the appearance of your product.

Dip coating

1) A metal is heated evenly in an oven and then plunged into fluidised powder (very fine plastic powder that's made to act like a liquid by passing gas through it) for a few seconds.

2) The metal, with this thin coating of plastic, is then put back in the oven and the plastic fuses (joins completely) to the surface.

3) It gives a soft, smooth finish so it's used for things like wire racks and tool handles.

Powder Coating

1) Plastic powder is sprayed onto the metal using an electrostatic gun.

2) The spray gun gives the powder an electrical charge. The particles repel all the others, since they've all got the same charge, so you get a very fine, even spray.

3) The object to be coated is given an opposite charge to the gun. This attracts the fine spray.

4) After spraying, the object is heated in an oven so that it undergoes a chemical change and sets hard.

5) This method gives an even coat and there's hardly any waste.

Galvanising

1) Iron rusts when it comes into contact with both oxygen and water, which are present in air.

2) Coating the iron with a barrier to keep out the water and oxygen can stop it rusting (oxidising). Another way is to place a more reactive metal such as zinc with the iron — water and oxygen then react with the zinc instead of with the iron.

3) Coating an object with zinc is known as galvanisation. The zinc layer protects the iron, but also if it's scratched, the zinc around the site of the scratch reacts first rather than the iron.

Polymers are Self-Finishing But You Can Decorate Them

1) Plastics don't need protective surface finishes because they're very resistant to corrosion and decay.

2) But for a nice appearance, you can use wet and dry paper to remove scratches, and follow that up with a mild abrasive polish or anti-static cream. Or, you could use a buffing machine.

3) Plastics can have decorative vinyl decals applied to them. Sheets of vinyl are cut to a design. The vinyl has an adhesive coating on the back so that it can be transferred to a plastic surface.

4) Designs can also be printed directly onto plastics. This can be done in a number of ways, for example, offset lithography (see p.54).

Bet you wish revision was self-finishing...

You may need to describe how to apply a finish or how to prepare a surface for finishing. So make sure you can.

Q1 What is the process used to treat timber to protect it from insect attacks and decay outdoors?

a) priming   b) tanalising   c) undercoating   d) galvanising   [1 mark]

Fabrics and Their Properties

Different materials have different <u>properties</u>. Flick back to <u>Section 2</u> if you need to refresh your memory.

The Properties of a Fabric Decide What It will be Used For

When deciding what fabric to use for a product, it's important to think about its <u>properties</u>.

Sportswear

Sportswear is often made from <u>polyester</u>, <u>elastane</u> or <u>polyamide</u> fibres. It needs to be <u>light</u> and <u>breathable</u>, and have other properties...

1) Polyester is <u>resistant to abrasion</u> — this means it can stand up to being used <u>intensively</u> (e.g. whilst exercising) again and again. It's also <u>strong</u> even when wet and <u>dries very quickly</u> — useful when being worn during exercise and when sweating.

2) Elastane is <u>really stretchy</u> so it's super flexible — great for when you're exercising. It's <u>not absorbent</u> and is usually used in <u>combination</u> with other <u>fibres</u> which are often able to <u>wick moisture</u> away from the body. Elastane is often used in <u>swimwear</u> and <u>cycling clothing</u>.

3) Polyamide fabric is made from a synthetic fibre. It's an <u>insulator</u> so can be very <u>warm</u> — useful for outdoor sportswear such as <u>ski-jackets</u>. It's also <u>easy to wash</u> (a very desirable quality for sweaty sportswear).

Furnishings

Furnishings are often made of <u>cotton</u>, <u>acrylic</u> or <u>wool</u> fibres. They need to be <u>soft</u> and <u>comfortable</u>, but also <u>durable</u>.

1) Cotton is used to make furnishings like cushions and curtains. It's <u>resistant to abrasion</u> — so it won't wear thin very easily, even when being sat on or handled every day.

2) Acrylic is <u>soft</u> and is quite <u>warm</u>, so will help to make your home nice and <u>cosy</u>. It's also <u>resistant to fading</u>, so will stay looking good for years. A <u>blend</u> of acrylic and other fibres is often used to <u>cover furniture</u> like sofas and chairs.

3) Wool is <u>strong and warm</u> — it's often used to make rugs and blankets.

In the 1900s, some swimsuits were made of wool. Hot, heavy and super absorbent — perfect for a sunny day on the beach...

Treatments can Change the Properties of Fabrics

1) <u>Chemical</u> treatments are applied to fabrics to <u>change their functional properties</u>. They are applied <u>during manufacture</u> at different stages of the production process. Here are a few for you to take a look at...

Flame retardance

1) <u>Flame retardant</u> treatments make fabrics <u>less likely to catch fire</u>.

2) They're often used on <u>flammable</u> fibres like <u>cotton</u>, and for specific products such as:

   • Work-wear for <u>welders</u> — e.g. overalls. Welders work with <u>flames</u> and <u>hot metal</u> so their clothing needs to be protected.

   • <u>Racing drivers</u>' overalls — their clothes need to be flame retardant in case they <u>crash</u>.

   • <u>Night clothes</u> (especially children's pyjamas). <u>Young</u> children's skin is more susceptible to <u>burns</u>, and they're <u>less aware</u> of the <u>risks</u> of <u>fire</u>, so it's important that their clothing is protected.

   • Fabric for <u>soft furnishings</u> — to make them meet <u>fire safety requirements</u>.

3) Using a fire retardant treatment on fabrics like cotton makes the fabric <u>slightly stiffer</u>, but the fabric is still <u>soft</u> and <u>cheap to produce</u>.

4) Some flame retardant treatments can be <u>washed out</u>, so care is needed when washing.

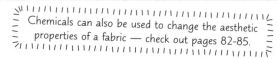

Chemicals can also be used to change the aesthetic properties of a fabric — check out pages 82-85.

Fabrics and Their Properties

Stain Protection

1) Fabrics can be made stain resistant by treating it with a mixture of silicone and fluorine compounds, or a Teflon® coating (this is also used on non-stick frying pans).
2) These treatments stop grease and dirt from penetrating the fabric.
3) Stain resistant treatments are used a lot on carpets and upholstery.
4) More recently, nanoparticles have been used in "self-cleaning fabrics" to give improved stain resistance (see p.32).

Rot Proofing

1) Mildew is a type of fungus — it can grow on fabrics made from natural fibres if they're kept in damp conditions.
2) If this is left untreated it can break down the fabric and cause it to rot.
3) Mildew growth can be reduced by applying a waterproof treatment such as PVC.

Some fabrics, like acrylic, are resistant to mildew.

Water-resistant Finishes

1) Chemicals (e.g. silicones) can be applied to the surface of fabrics to stop water droplets passing through.
2) These finishes don't make the fabric waterproof — if the surface becomes saturated (completely covered in water) the water will leak through.
3) Fabrics with a water-resistant finish can be washed and dry-cleaned without affecting the performance of the finish.
4) Nylon is often given a water-resistant finish and used to make coats and tents.

Water forms beads on the surface of the fabric.

2) The properties of a fabric aren't only changed by chemical treatments though — how the fabric has been constructed can also have an impact...

Laminated Fabrics are Constructed from Different Layers

1) Laminated fabrics are made up of two or more different layers that are stuck together.
2) Lamination can add useful properties, e.g. a layer of foam could be used to make a fabric more insulating. It can also be used with delicate fabrics — to add strength from an under layer, or protection from an outer layer.

For other ways of adding strength to fabric see page 37.

3) Waterproof coats are made from laminated fabrics. They have to be warm, breathable and waterproof, so are often made of a layer of natural or synthetic fibres that is then laminated with a waterproof layer such as rubber or PVC.
4) Some laminated fabrics in waterproof coats also use a membrane layer that allows sweat to escape but doesn't let rain in. This helps to keep the wearer warm and dry.

I don't think my brain has the properties needed for revision...

So, different fabrics have properties that make them useful for different things. And properties can be added to a fabric to make it safer or more useful. Wonderful stuff. Now have a go at the questions below.

Q1 Name two properties of polyamide fabric which make it suitable for use in sportswear. [2 marks]

Q2 State a property of cotton that makes it suitable for use in cushions.
Give a reason for your answer. [2 marks]

Q3 Give an example application of a flame retardant treatment. [1 mark]

Q4 What is a laminated fabric? [1 mark]

Standard Components and Tools

Next up, you need to know about how fabric is sold and a few standard components you can use with fabrics.

Fabric is Sold in Standard Widths

When you're buying fabric you normally just choose the length that you want
— the width is determined by the width of the roll that you buy it from.

1) Fabric comes on rolls of standard widths,
   e.g. 90 cm, 115 cm and 150 cm. So you
   only need to specify the length you require.

2) Fabrics also come in different weights.
   The weight of a fabric depends on
   how tightly the fabric is knitted or
   woven, and on the thickness of the
   yarn used (i.e. 1-ply, 2-ply etc.).

EXAMPLE:

A 4.5 m length of fabric is cut from a 115 cm-wide
roll. Calculate the area of this piece of fabric.

First convert the lengths to the same unit (e.g. metres).
1 m = 100 cm, so 115 cm = 115 ÷ 100 = 1.15 m

Then calculate the area by multiplying length by width.
$$= 4.5 \text{ m} \times 1.15 \text{ m}$$
$$= 5.18 \text{ m}^2 \text{ (to 2 decimal places)}$$

Standard Components include Fastenings

Standard components are the bits and pieces that you use in addition to the fabric to make a textile product.
They can be functional, e.g. fastenings like zips to close your coat, or decorative, e.g. additional buttons.

Zips

1) They can be made out of plastic or metal,
   and can be big and bulky or small and
   concealed (hidden) in your textile design.

2) Some zips are fixed at one end (e.g. on
   handbags). Zips on jackets are not fixed
   (so you can get the jacket off).

3) Zips with two sliders can be opened in two
   directions — the ends can be fixed (e.g. on
   suitcases) or open (e.g. on some jackets).

Advantages
1) They're a secure fastening — they close the
   product fully with no gaps.
2) They're quick to do up and undo.
3) They're quick and simple to attach.
4) They lay flat and don't add bulk.
5) They're hard-wearing and can be washed.
6) Colours can match or contrast with the fabric.

A **disadvantage** is that they can snag delicate fabrics.

Toggles and Buttons

1) These are sewn on and require a
   buttonhole or a loop to fasten to.

2) They can be made of any hard material
   — plastic, metal, wood, and even glass.

Advantages
1) They're easy to attach and replace.
2) Colours can match or contrast with the fabric.
   (Buttons can be fabric-covered to blend in.)

Disadvantages
1) They can fall off — a choking
   hazard on children's products.
2) They can be damaged in
   the wash.

Velcro®

1) Velcro® comes in two halves —
   a rough tape and a smooth tape.
   Nylon hooks on the rough half attach
   to soft loops on the smooth half.

2) Velcro® requires a large amount of force to open
   it — so it needs to be firmly attached to the fabric
   and isn't suitable for using with delicate fabrics.

Advantages
1) Safe and soft (good for
   children's products).
2) It can be machine washed.
3) It's hard-wearing.

Disadvantages
1) Hooks collect fibres
   over time and
   become less sticky.
2) Not very decorative.

Press studs (or poppers)

These can be used to fasten items that need to be opened and closed quickly. They can be made
of metal or plastic and come in different sizes depending on the strength of fastening you require (bigger
metal press studs are harder to open than small plastic press studs). And they're not very decorative.

Standard Components and Tools

You also need to know about <u>tools</u> that can be used when <u>making textiles products</u>...

Use the Right Tools for Each Task

For Cutting

1) Use <u>PAPER SCISSORS</u> to cut out <u>patterns</u> (take a look at p.43).

2) Use <u>DRESSMAKING SCISSORS</u> (also called <u>fabric shears</u>) to cut <u>fabric</u>. These have long, very sharp blades that cut through fabric more easily and neatly.

3) Use <u>EMBROIDERY SCISSORS</u> for more <u>delicate</u> jobs, e.g. snipping threads, or clipping curved seams to help press them. They have short, sharp blades.

4) Use <u>PINKING SHEARS</u> to cut fabric with a <u>zigzag edge</u> — this helps prevent fabric from fraying.

Scissors and shears use shear forces to cut materials — see p.36

Use <u>CRAFT KNIVES</u> to cut <u>stencils</u> (e.g. if you want to <u>spray</u> a <u>design</u> onto fabric). You'll do a <u>neater</u> job than using scissors.

Use <u>SEAM RIPPERS</u> (quick unpicks) to <u>unpick</u> <u>seams</u>. Doing it by hand is slower, and scissors might accidentally cut the fabric.

For Sewing

1) You can use <u>PINS</u> to <u>hold</u> the fabric together before stitching with a sewing machine.

2) They help keep your fingers <u>away from the needle</u> when you feed fabric through (be careful not to catch them with the machine though).

1) Use <u>NEEDLES</u> for <u>hand</u> stitches, e.g. <u>embroidery</u> stitches, attaching <u>beads</u> to fabric, or <u>tacking stitches</u>.

2) Use a needle that's the <u>right size</u> for the <u>thickness</u> of the <u>fabric</u> and the <u>thread</u> you're using.

There's more about pinning, tacking and using sewing machines on the next page.

For Measuring and Marking out

Use flexible <u>MEASURING TAPES</u> to accurately <u>follow</u> <u>curved</u> surfaces.

You can also use <u>TAILOR'S CHALK</u> and <u>PATTERN MASTERS</u> to help draw and mark out <u>patterns</u> (see p.116)

For Pressing

<u>DRY IRONS</u> use <u>heat and pressure</u> to press creases out of the fabric and flatten seams. <u>STEAM IRONS</u> are more <u>effective</u> — they use <u>water and steam</u> as well as heat and pressure.

Irons can also be used to <u>apply designs</u> from <u>transfers</u> onto fabric (<u>heating</u> the transfer causes the design to <u>imprint</u> onto the fabric), or to <u>fix</u> designs done with <u>fabric crayons</u> and pens.

Standard components might pop up in the exam...

...so you should make sure you remember all the different types. Make sure that you've understood everything, then have a go at the questions below. If you're struggling, read over these pages one more time, then try again.

Q1   A fabric is sold from a roll that is 90 cm wide. Calculate the surface area of a 3 m length.   [2 marks]

Q2   Give one advantage and one disadvantage of each of the following components:

    a) zips   b) Velcro®   [4 marks]

Q3   Which type of scissors cut a fabric in a way that helps prevent it from fraying? Explain how.   [2 marks]

Joining and Shaping Fabrics

There are loads of different types of <u>machinery</u> that can be used to make <u>joining</u> and <u>shaping</u> fabrics <u>quicker</u> and <u>easier</u> to do. We'll cover these in a bit, but first off, we should talk about <u>methods</u> that are done <u>by hand</u>...

Pinning or Tacking Joins Fabrics Together Temporarily

Fabrics can <u>slip</u> when you're trying to machine or hand sew them together, so it's best to <u>pin</u> or <u>tack</u> them in place first.

line you're going to sew along

Pinning

1) Pinning involves putting <u>pins</u> at <u>right angles</u> to the edge of the fabric.
2) If you're machine sewing, <u>remove</u> the pins as you come to them.

Tacking

1) Tacking (sometimes called <u>basting</u>) involves <u>hand sewing</u> long running stitches (about 1 cm). They should be in a <u>different colour</u> to the fabric, so you can easily see them.
2) Tacking holds the fabric together more <u>securely</u> than pinning (it's best to pin before tacking though).
3) Once you've tacked, remove all the <u>pins</u>, then you can see what your pieces will look like when joined. You can then <u>stitch</u> over your tacking.
4) Tacking stitches are <u>temporary</u> — they'll need to be <u>removed</u> once the stitching has been completed.

If your fabric will be <u>damaged</u> by pinholes or by tacks, e.g. PVC-coated or leather fabrics, you could use <u>paperclips</u> or <u>bulldog clips</u> to hold it in place instead.

Sewing Joins Fabrics Permanently

1) Once you've finished pinning or tacking your fabrics together, you're ready to start <u>sewing</u>.

2) <u>Hand-sewing</u> is great for <u>small, quick tasks</u> or those that need <u>precision</u>, such as embroidery or darning. For <u>bigger projects</u> it's quicker to use a <u>sewing machine</u>. They <u>speed up sewing</u>, and produce <u>neat</u>, <u>even</u> stitches for a <u>high-quality finish</u>.

3) Most sewing machines use two threads — one on a <u>bobbin</u> under the sewing plate, the other (the <u>top thread</u>) on the spool pin on top of the machine. The machine <u>interlocks</u> the two threads to make <u>stitches</u>.

1) <u>Before you start</u>, choose the <u>right needle</u> for your fabric and thread thickness and <u>fasten</u> it securely into the machine.

2) It's a <u>good idea</u> to do some lines of stitching on a <u>small sample</u> of your fabric first, so you can <u>check</u>:
   - That the thread <u>tension</u> is right. If the tension of the two threads is <u>balanced</u> then you'll get an <u>even stitch</u>, which <u>isn't too tight</u> or <u>too loose</u>.
   - That the stitch <u>type</u> and stitch <u>length</u> are correct.

tension control
spool pin
bobbin winder
take-up lever
balance wheel
thread guide
stitch selector
presser foot
needle
presser foot lifter
sewing plate
stitch length control
bobbin & bobbin case

A variety of different glues can also be used to permanently join fabrics.

Who knew revision could be sew easy...

There are a couple things you can do to make sure your fabrics are neatly sewn together. You should pin or tack your fabrics to make sure they don't move when sewing them together. If you're using a sewing machine, you should test it with a spare bit of fabric first — this'll check you've set it up suitably for the fabric you're joining.

Joining and Shaping Fabrics

Overlockers can Sew Seams and Finish Edges at the Same Time

1) Overlockers are used to <u>finish edges</u> to <u>stop</u> them from <u>fraying</u>. They do this by <u>enclosing</u> the edge, or edges, in a thread casing.

2) An overlocker works by using several top threads, but no bobbin. It also has a <u>blade</u> to <u>trim</u> the fabric edge before it's enclosed.

3) They can be used just to finish seam edges, or to <u>sew, trim and neaten</u> the seam all in <u>one go</u>.

4) They can be used for <u>side seams</u> in <u>stretchy</u> clothes like T-shirts.

Three different fabrics with overlockers.

CAM can be Used to Cut and Sew Fabrics

CAM is covered in more detail on page 4 if you need to go over what it is. It has lots of <u>different uses</u> within the <u>textiles industry</u>, from embroidery to cutting fabric...

Embroidery and Knitting Machines

1) <u>CAM embroidery machines</u> use <u>CAD</u> data to sew designs. They have many needles that change <u>automatically</u> as different coloured threads are needed.

2) <u>CAM knitting machines</u> use <u>CAD</u> data to control the <u>stitch pattern</u> and other design features. They can produce rolls of knitted fabric or even whole one-part garments.

3) <u>Computer control</u> makes these machines <u>very fast</u> and <u>accurate</u>, and the CAD <u>data</u> can be <u>quickly changed</u> to produce <u>new products</u>.

Cutting Machines

1) Before cutting, fabric is <u>automatically spread out</u> on the cutting table in <u>layers</u>.

2) A <u>CAM cutting machine</u> automatically cuts out the fabric pieces, following <u>CAD</u> lay plan instructions. (A lay plan is a plan that shows you how to position pattern pieces on fabric in the most efficient way.)

3) The machine cuts through <u>all the layers</u> at once, which makes the cutting process <u>really quick</u>.

4) It cuts the fabric <u>accurately</u> at <u>high speed</u> using vertical knives, high-pressure water jets or lasers (see page 5).

You can cut fabric by hand too using scissors and shears — see p.77.

Sewing Machines

1) <u>Industrial</u> sewing machines are very strong as they need to work at <u>high speeds</u>.

2) <u>CAM</u> sewing machines are used to carry out certain processes <u>automatically</u> — e.g. sewing buttonholes and attaching pockets.

The overlocker — try getting out of that one Mr Houdini...

Q1 What is a tacking stitch used for? [1 mark]

Q2 Describe how a sewing machine forms stitches. [3 marks]

Q3 Why is an overlocker used to finish edges? [1 mark]

Q4 a) Give an example of one process that a CAM sewing machine could carry out automatically. [1 mark]

b) Suggest another piece of textiles equipment that uses CAM. [1 mark]

Joining and Shaping Fabrics

Now you've got to grips with the cutting and sewing basics, these pages cover the different types of <u>seams</u> and a few techniques you can use to <u>add shape</u> and <u>detail</u> to your work...

A Seam is where Two Pieces of Fabric are Joined Together

1) <u>Seams</u> are held together with <u>stitches</u>. They need to hold fabric <u>securely</u> and be <u>strong</u> enough to stand up to the <u>strains</u> put on the product.

2) There are different types you can use, depending on the <u>fabric</u> and what the product will be used for.

Plain (flat) seam — the simplest to do

1) <u>Plain</u> seams look <u>neat</u> on the outside — you can only see a thin joining line.

2) They are used for fabrics which <u>aren't</u> going to be under <u>too much strain</u>.

3) If your fabric is <u>stretchy</u>, use <u>stretch or zigzag</u> stitch to allow the seam to stretch.

A seam allowance is the extra material between the seam and the edge of the fabric — it's usually about 1.5 cm wide.

1) To make a <u>plain</u> seam, arrange two pieces of fabric so their outward-facing sides are touching. Then <u>pin</u> or <u>tack</u> to hold the fabric in place.

line of stitching

2) Stitch about <u>1.5 cm</u> in from the edge of the fabric (patterns have a <u>seam allowance</u> to allow extra fabric for this).

3) <u>Strengthen</u> the seam by <u>reversing</u> back over it for a few centimetres.

4) <u>Finish</u> the seam edges to <u>stop</u> them <u>fraying</u>. Then <u>open</u> out the seam and <u>iron</u> so it <u>lies flat</u>.

French seam — encloses the seam edges

1) French seams <u>enclose</u> the <u>raw edges</u> — they're used for <u>fine fabrics</u>, or fabrics which are <u>likely to fray</u>.

2) They're <u>strong</u>, but <u>not bulky</u>.

3) They're good for <u>baby clothes</u> — there are no rough edges to irritate the skin.

Flat-felled seam — strong and durable

1) Flat-felled seams have <u>two lines</u> of <u>stitching</u>. A plain seam is sewn and the <u>seam edges</u> are <u>enclosed</u> by wrapping one edge around the other. The whole thing is <u>ironed flat</u>, then a second line is stitched on top.

one seam tucked under the other

lines of stitching

2) They <u>stop</u> edges from <u>fraying</u> and their <u>strength</u> means they're used in hard-wearing clothes like <u>jeans</u>. You <u>wouldn't</u> use them with <u>delicate</u> fabrics because of the extra <u>bulk</u>.

Piping — adds definition to a seam

1) <u>Piping</u> can be used at <u>seams</u> to add <u>decoration</u> or to <u>strengthen</u> a product. It <u>stands out</u> from the seam, adding <u>definition</u>.

2) Piping is made from a <u>strip of material</u> that has <u>folded fabric</u>, or a <u>cord</u>, sewn inside.

3) It often has a bit of <u>spare fabric either side</u> so that it can be attached to a seam.

4) To add piping, <u>line up</u> your pieces of fabric on either side of the piping's seam allowance. Make sure that you lay your fabric <u>face down</u> onto the piping's seam allowance (so that you won't be able to see the seam once it's finished). Sew a <u>line of stitching over the fabric</u> — you can sew both pieces of fabric to the piping at once.

5) Unfold your fabric and <u>iron the underside of the seam</u> — this will flatten the excess material and make it look neater.

A section of piping

fabric

row of stitching

cord

Joining and Shaping Fabrics

Quilting uses Wadding

1) <u>Wadding</u> is any soft material that is used to <u>stuff</u> or <u>line</u> something.

2) <u>Quilting</u> uses <u>wadding</u> between <u>two layers</u> of fabric which are then <u>stitched together</u> in <u>straight lines</u> or in a <u>pattern</u>.

3) Quilting is often used to give added <u>warmth</u> to a product (e.g. an anorak or bed cover). The wadding <u>traps warm air</u> between the <u>layers</u> of fabric.

Advantages:
1) Can create interesting <u>3D effects</u>.
2) Gives <u>warmth</u>.

Disadvantages:
1) Requires a <u>lot</u> of <u>material</u>.
2) <u>Time consuming</u>.

Gathering and Pleating are Shaping Techniques

You need to know some different techniques for <u>shaping</u> fabrics. Both <u>gathering</u> and <u>pleating</u> use <u>excess material</u> to create <u>detail</u>. They can also have <u>practical uses</u> — gathering can be used to give a <u>better fit</u>.

Gathers Pull in the Fabric Evenly along its Length

<u>Gathers</u> can be used at <u>waistbands</u> or <u>cuffs</u>.

1) <u>Knot</u> the threads at the start, or if machine sewing, leave a long thread that can be <u>wrapped around a pin</u>.

2) Sew <u>two parallel rows</u> of stitches in the <u>seam allowance</u>. Use <u>small running stitches</u> when sewing by hand. If you're using a sewing machine, set the machine to its <u>longest stitch</u> and <u>loosest tension</u>.

3) <u>Pull the threads</u> and ease the fabric along until it is drawn to the size you want.

4) If you want to <u>fix the gathering in place</u> (e.g. if you're making a skirt), knot the threads at the <u>end of the gathered fabric</u>.

Pleats are Folds in Fabric

<u>Pleats</u> can be used to <u>shape</u> skirts.

1) Allow <u>three times</u> the finished width of fabric and <u>mark</u> the position of the pleats at <u>even widths</u> across the fabric.

pleats

2) <u>Fold</u> the pleats, <u>pin</u> them into position and stitch across the <u>top</u> of them to keep them in place.

3) <u>Pressing</u> gives the pleats <u>sharp creases</u>.

A pleated skirt

Let me just gather my thoughts...

Lots of different techniques for joining fabrics on these pages — fascinating stuff. Combine that with your new knowledge of gathering and pleating and your prototype will be shaped and formed to perfection in no time.

Q1 State which types of seam would be most suitable in the following items of clothing and explain why.

a) baby clothes [2 marks]

b) jeans [2 marks]

Q2 John wants to add decoration to his sofa cushions. Suggest a technique he could use to do this. [1 mark]

Q3 Give an example of where a gather can be used in clothing. [1 mark]

Q4 Give one advantage and one disadvantage of quilting. [2 marks]

Dyeing

Dyeing <u>changes the appearance</u> of a fabric — it can be used to <u>add colour</u> and <u>pattern</u>, and there's a few different methods you could use. Sometimes it even involves tea and onions — read on to find out how...

There are Natural Dyes and Chemical Dyes

Natural Dyes

1) Until the 1850s all dyes came from <u>natural sources</u>.
2) They're made from things like <u>onions</u>, <u>beetroot</u>, <u>tea</u>, <u>raspberries</u> or <u>flowers</u> etc.

Fibres and fabrics can be dyed at <u>different stages</u> of the manufacturing process — individual <u>fibres</u>, <u>yarns</u> or whole pieces of <u>fabric</u> can all be dyed. The methods here are for <u>dyeing fabrics</u>.

Chemical Dyes

These were invented in the <u>1850s</u>.

<u>Advantages</u>
- Colours are <u>brighter</u>.
- <u>Easier</u> and <u>cheaper</u> to make.
- Exactly the same colour can be achieved <u>repeatedly</u>.

<u>Disadvantage</u>
Some are <u>toxic</u> — they can be <u>harmful</u> to <u>people</u> and the <u>environment</u>.

Some Fabrics are Better for Dyeing than Others

1) <u>Natural</u> fibres (like cotton, wool and silk) are the <u>best</u> for dyeing as they're very <u>absorbent</u>.
2) The <u>colour of the fabric</u> you begin with makes a difference to the final colour — for example if you dye <u>white fabric red</u>, you get <u>red fabric</u> — if you dye <u>yellow fabric red</u>, you get <u>orange</u>.
3) Fabrics that have an <u>uneven colour</u> need to be <u>bleached</u> before dyeing to ensure an even colour.
4) For some fabrics and dyes you need to use a chemical called a <u>mordant</u> (e.g. salt) to <u>fix</u> the colour to the fabric. This makes the fabric <u>colour-fast</u> — the dye won't come out in the wash.

Batch Dyeing is a Commercial Dyeing Technique

1) Commercial dyeing involves dyeing a <u>huge amount</u> of fabric at a time. The fabric is dyed a <u>uniform colour</u>. It can be done <u>continuously</u> — this is where <u>very long lengths</u> of fabric are dyed the <u>same</u> colour in a continuous process. Or it can be done in <u>batches</u>...
2) In <u>batch dyeing</u>, a <u>batch</u> of fabric (a specific amount) is dyed with <u>one colour</u>, then another batch with a different colour, and so on. Here's one way it's done:

1) The fabrics are initially produced and <u>stored without dyeing</u>.

2) When a <u>batch</u> of coloured fabric is needed, the <u>required amount</u> is dyed.

Batch of undyed fabric on roller

Dyed fabric collects on a second roller

Dye vat

Fabric passes back and forth between rollers

3) The fabric is <u>mounted on two rollers</u> and passed <u>back and forth</u> through the dye until all the dye is <u>used up</u>.

4) The dyed material is put in a separate machine to <u>fix</u> the dye (to stop the colour from running and to <u>wash off</u> excess dye).

During the fixing process, dye is often oxidised (it gains oxygen). This can be done by exposing the dye to oxygen in the air or using a specific chemical. Oxidation causes the dye to become insoluble, so it won't dissolve in water and come out in the wash.

Colours can go out of fashion quickly.

3) Batch dyeing is useful in industry, as it allows textiles <u>manufacturers</u> to <u>respond quickly</u> to orders for a <u>specific colour</u> of fabric.

Dyeing

Hand Dyeing — Use a Resist to make Patterns

One of the advantages of hand dyeing is that you can add designs to the fabric using a 'resist'. A 'resist' is something which prevents the dye from reaching the fabric — it's applied in a pattern before dyeing.

Tie-Dye is Easy to do

Different ties give different patterns

1) Fabric is tied with string or rubber bands to create a resist.
2) The fabric is then immersed in dye.
3) The dye doesn't get to the tied areas.
4) Once the dye has dried and the fabric is untied, the pattern is revealed.

Small circles are formed when pebbles or buttons are tied in.

Concertina pleats give parallel lines.

Tying the fabric out from the centre produces a 'sunburst' effect.

This vest has been tie-dyed.

Advantages
1) Every piece will be unique.
2) The equipment is cheap and readily available.

Disadvantages
1) The outcome is unpredictable.
2) You can't repeat a pattern exactly.
3) You can't create a detailed pattern.
4) It's time-consuming for large areas.

Batik — Good for more Detailed Designs

In batik the resist used is usually hot wax.

1) The fabric is stretched across a frame. Then the hot wax is applied with a brush or a tjanting (pointed tool for dripping the wax) to create a pattern.

2) Once the wax is dry, the dye is painted on, or the fabric can be immersed in a dye bath.

3) The wax is ironed off to reveal the pattern.

Advantages
1) It's a more precise way of adding patterns to fabrics than tie-dye.
2) Patterns can be more detailed.
3) Every product will be unique.

Disadvantages
1) It's time-consuming — each part of the design has to be painted separately.
2) You need to be careful when working with hot wax, and it's easily dripped in the wrong place.
3) It can be tricky to iron all the wax out.

Batik works best on natural fabrics — they absorb hot wax and cold dye more easily than synthetics do.

Finished this page alive — you'll live to dye another day...

Dye made from onions — no wonder it smells so bad. Make sure you remember the different dyeing methods that we've covered on these pages, and the ways that detail can be added — it might come up in the exam.

Q1 Why might a chemical dye be chosen over a natural one? Give two reasons. [2 marks]

Q2 What are mordants used for? Give an example of a mordant. [2 marks]

Q3 Name a process that allows large quantities of fabric to be dyed at once. [1 mark]

Q4 What type of resist might you use in batik? [1 mark]

Printing

Just like with dyeing, there are different methods of printing that can be done <u>by hand</u> or <u>machine</u>. You'll need to understand <u>how they work</u>, and why you might use each type. Lots to learn. Here we go...

Printing is Used to Apply a Design to Fabric

1) Printing is the process of applying <u>ink</u>, <u>dye</u> or <u>paint</u> to fabric in <u>defined patterns</u>.
2) Printing can be done in <u>small quantities</u> by <u>hand</u>, or in <u>large quantities commercially</u>.
3) Materials with a <u>tight weave</u> are best for printing on because they have a <u>smooth surface</u> for the dye to be applied to and <u>no surface pattern</u> to <u>detract</u> from the design.

A printed dress

To Print by Hand — Use Block Printing...

You need a <u>printing block</u> with a <u>raised design</u>.

1) You can make one by <u>drawing</u> a pattern on a piece of wood and then <u>cutting</u> the <u>background away</u>, leaving the design <u>raised</u>.
2) Or you can <u>stick</u> pieces of <u>card</u> or <u>string</u> onto a solid <u>block</u>.
3) You can also <u>buy rubber blocks</u> with designs already on them.

The Printing Process...

1) <u>Printing ink</u> is applied to the <u>raised surface</u> of the block.

2) The block is then <u>pressed down</u> onto the fabric...

3) ...leaving a <u>reversed</u> image on the fabric.

Advantages

1) You can print with <u>different colours</u> using the <u>same block</u>.
2) You can use <u>several blocks</u> to build up a more <u>complicated</u> design.
3) You can easily <u>repeat</u> designs.
4) You can use the same block <u>many</u> times before it <u>wears out</u>.

Disadvantages

1) <u>Making</u> the blocks takes a <u>long time</u>.
2) It's not good for <u>fine detail</u>.

In industry, <u>engraved rollers</u> are used instead of blocks. The rollers are inked as the fabric is <u>continuously</u> run under them. The rollers are <u>expensive</u> to make, so it's only cost-effective for large amounts of fabric.

...or Flat-Bed Screen Printing

1) The <u>screen</u> is a <u>frame</u> with <u>fine mesh</u> covering it.
2) A <u>stencil</u> is cut from card or acetate by hand or using CAD/CAM and put <u>beneath</u> the screen (on top of the fabric). Alternatively, parts of the screen itself can be <u>blocked off</u> to create the design.
3) Printing <u>ink</u> is poured onto the screen.
4) A <u>squeegee</u> is pressed down and drawn across the screen, forcing the ink <u>through</u> the mesh and the holes in the stencil.
5) The screen is <u>lifted up</u> and the design is left on the fabric.

Advantages

1) Can produce <u>intricate patterns</u> that can be <u>repeated accurately</u>.
2) Good for printing <u>large areas</u> of colour.
3) Easy to use <u>many colours</u> — by using several screens.
4) Printing is <u>quick</u> to do (once you have the screens and stencils).

Disadvantages

1) <u>Making</u> the <u>screens</u> and <u>stencils</u> takes a long <u>time</u>.
2) Each colour has to be applied <u>separately</u>.

The Squee Gees

<u>Flat-bed Screen</u> printing by <u>hand</u>...

Squeegee

Ink

Mesh

Screen printing can also be done by a machine — see the next page.

Printing

Screen Printing is done Industrially with Machines

1) Industrial screen printing uses <u>machines</u> to repeat the <u>same pattern</u> all the way along a <u>long length</u> of fabric.

2) There are <u>two</u> methods of industrial screen printing. Like screen printing by hand, they both use some sort of <u>screen</u> and a <u>squeegee</u>.

The screens can be made using CAD and CAM.

Flat-bed screen printing

1) This is similar to flat-bed screen printing <u>by hand</u>. <u>Several</u> screens are used — one for <u>each</u> colour.

2) The fabric passes <u>under</u> the screens on a <u>conveyor belt</u> and the colours are applied one after the other.

Screens

Fabric Reel

The squeegees are <u>metal rods</u> that move backwards and forwards across the screen.

Between each ink application, the screens are <u>lifted</u> and the fabric <u>moves along</u> the belt to the next screen position.

At the end of the conveyor belt the colour is <u>fixed</u> to stop it running.

Rotary screen printing

1) Rotary screen printing is similar to flat-bed screen printing but the <u>screens</u> are on <u>cylinders</u>.

2) This is a really <u>fast</u> and <u>widely used</u> industrial method of printing.

3) <u>300 m</u> of fabric can be printed every <u>minute</u>.

Rotary screens

A <u>different colour dye</u> is pumped into each of the rotary screens.

The dye is squeezed through holes in the <u>rotary screen</u> by a <u>cylindrical squeegee</u>.

<u>Side-on</u> view of <u>rotary screen</u>:

fine wire mesh and screen.

dye

squeegee

conveyor belt

Machines can also print on individual garments, e.g. T-shirts.

Advantages

1) Just as with <u>hand</u> screen printing, <u>intricate patterns</u> using <u>many colours</u> can be <u>repeated</u> accurately.

2) The <u>machinery</u> makes the printing process <u>much quicker</u> and <u>large amounts</u> of fabric can be printed at once. Rotary printing is the <u>quickest</u> way — as fabric is passed along the conveyor belt <u>continuously</u>.

3) The <u>screen making</u> process can be <u>computer-controlled</u>, which makes it much <u>quicker</u>.

Disadvantages

1) Setting up commercial <u>machinery</u> is <u>expensive</u>.

2) <u>Screen making</u> takes a <u>long time</u>, unless it's done using CAD/CAM.

Printing sending you to sleep? Just go and find a flat-bed...

Right, there's a lot of information over these pages but it's important to learn it — remember to think about the advantages and disadvantages of each process too — you might be asked questions about it in the exam.

Q1   Why are materials with a tight weave better for printing on? [2 marks]

Q2   Sophie wants to block-print some letters onto her fabric.
Describe one way in which she could make her own printing block. [1 mark]

Q3   What is the squeegee used for in flat-bed screen printing? [1 mark]

Q4   Give one advantage and one disadvantage of using industrial screen-printing techniques. [2 marks]

Properties of Components in Systems

Electronic and mechanical components are made from materials that have specific properties.

Systems are Made from Materials with Suitable Properties

1) Mechanical systems are made up of lots of individual physical parts. In much the same way, electronic systems contain many electrical components. Each one has a particular function in the system.

2) Different components in a system are commonly made from different materials. This is because we often require materials with a certain set of properties to perform a specific function.

Motor Vehicles

1) Steel is often used for vehicle parts, e.g. the brake discs, engine, wheels, and drive shaft (a rod that transmits rotary motion and power from the engine to the wheels). It's strong, easy to work, malleable, and cheap. These properties mean that durable vehicle parts can be made relatively cheaply and easily.

2) Car bodies are also made from steel because it's tough. In an accident it's designed to crumple, absorbing most of the impact — this helps protect the passengers.

3) Aluminium is a more corrosion-resistant alternative to steel for car body parts. It's also less dense, so aluminium parts are lighter than steel — this improves performance (e.g. acceleration and fuel economy). It is more expensive and harder to work though, so it's more commonly used in expensive vehicles.

4) Strips of electrically conductive material on the inside of a car's rear window can be used to defog it. When an electric current is passed through the strips, they heat to around 30 °C and clear the glass.

5) Engines get hot when in use so engine electronics are made from materials that can cope with high temperatures. These electronics have to be able to work in freezing conditions too.

Domestic Appliances

1) Washing machines use a drive belt — a band that transmits rotary motion from the motor to the spinning drum. They're often made from rubber or a polyurethane plastic, which is:
- Strong — able to resist pulling forces without breaking.
- Flexible so the belt can bend.

The spinning drum, and any pumps and valves, must be resistant to corrosion as they're exposed to water during use — stainless steel (see p.18) or plastic is often used.

2) A heating element is a wire that gets hot when a current passes through — this is caused by the electrical resistance of the element. They can be found in appliances such as toasters, kettles and electric ovens.

3) Heating elements are often made of a nickel-chromium alloy. It has a high melting point (so it can get hot without melting) and is corrosion-resistant (so it can cope with being exposed to air and water).

Properties of Materials can be Modified for Specific Purposes

Photosensitive Material in Printed Circuit Boards

1) Photosensitive materials respond to being exposed to light. For example:
- The photographic film in a camera goes through a chemical change that is dependent on the amount of light present. This can be developed into a photo.
- A light-dependent resistor contains a material with an electrical resistance that varies with light levels (see p.25).

2) When making printed circuit boards (PCBs), a process called photo-etching is used to transfer the design of the circuit onto the board.

3) The circuit board initially has a photosensitive layer on top of a layer of copper. When this layer is exposed to UV light it is modified — this allows copper to be removed from specific bits of the board (see p.92).

Don't open the curtains! I'm feeling photosensitive...

Properties of Components in Systems

Anodised Aluminium

1) Some metals (such as aluminium) naturally <u>react</u> with <u>oxygen</u> in the <u>air</u> to form a <u>layer</u> of <u>metal oxide</u> on <u>exposed surfaces</u> — this is known as <u>corrosion</u> or <u>oxidation</u>.

2) When aluminium oxidises, it forms <u>aluminium oxide</u>. Unlike rust, this is a <u>hard</u> material that <u>sticks</u> strongly to the fresh aluminium beneath, <u>protecting</u> it from <u>further corrosion</u>.

3) <u>Anodisation</u> is a process that <u>thickens</u> the <u>protective</u> oxide layer that is already present using an <u>electric current</u>. In <u>aluminium</u>, this makes the surface <u>harder</u> (and even more <u>resistant to corrosion</u>).

4) Anodised aluminium is used to make loads of stuff including <u>aircraft parts</u>, <u>pans</u> and <u>window frames</u>.

Iron and steel oxidise in the presence of air and water. This causes rust to form on the surface of the metal.

Anodised aluminium pans have a hardened surface that's difficult to scratch. This makes them durable and means you shouldn't get aluminium in your food.

Current and Voltage Ratings Prevent Damage to Components

1) Most <u>electronic components</u> have a <u>current</u> and <u>voltage rating</u> — these are the <u>current</u> and <u>voltage values</u> that the component is <u>designed</u> to <u>work at</u>.

2) They're <u>set</u> by the <u>manufacturer</u> and should be <u>followed</u> to <u>avoid damage</u> being done to the component.

Remember, the voltage from a power supply pushes an electric current around a circuit.

mA stands for milliamp — 1 mA is the same as one thousandth of an amp (0.001 A).

For example, a <u>light-emitting diode</u> (LED) will typically have a current rating of around <u>20 mA</u> (0.02 A) and a voltage rating of around <u>2 V</u>. If the current and voltage are a lot <u>smaller</u> than this (e.g. 2 mA or 1 V), the LED may be <u>dim</u> or <u>not light up at all</u>. However, if we <u>go much over</u> the ratings (e.g. 1 A and 10 V), the LED will <u>heat up</u> and become <u>damaged</u>.

3) Components should be <u>bought</u> to <u>match</u> a chosen <u>power supply</u>. For example, <u>mains</u> power supplies a <u>high voltage</u> (<u>230 V</u>), so is often used for <u>power-hungry</u> components. <u>Batteries</u> commonly have a <u>lower voltage</u> than mains power and are usually used in <u>smaller</u> or <u>portable</u> products.

Battery Type	Voltage Supplied	Applications
AA or AAA	1.5 V	Alarm clocks, remote controls, torches
PP3 (9-volt battery)	9 V	Smoke detectors, walkie talkies, radios, microphones
Button cells	1.5 - 3 V	Watches

4) <u>Resistors</u> are often used in circuits to <u>modify</u> the <u>current</u> from a power supply (see p.24) — this makes sure components are operating at their <u>current rating</u> and will not be <u>damaged</u>.

My current rating is pretty low — I don't even like raisins...

Corrosion isn't all bad — anodisation is used to thicken the corroded layer of a metal to make its surface harder...

Q1   Explain why using aluminium in motor vehicles can be better than using steel. Your answer should include two different properties of aluminium.   [4 marks]

Q2   Suggest one property of rubber that makes it useful in washing machine drive belts.   [1 mark]

Q3   Name the process that makes the surface of aluminium harder.   [1 mark]

Standard Components in Systems

There are loads of <u>standard components</u> in <u>electronic</u> and <u>mechanical systems</u>. We'll cover just a <u>few</u> here...

A Common Set of Resistors is the E12 Series

1) <u>Resistors reduce the current</u> flowing round a circuit. E12 resistors are <u>fixed resistors</u> — this means they have a <u>constant resistance</u>, unlike LDRs and thermistors (see p.25).

Resistance is how much a component <u>limits</u> the <u>current</u> flowing through it.

2) Manufacturers don't make resistors for <u>every</u> single possible <u>resistance</u> that a customer could want — instead, they make different <u>series</u> of resistors, which offer different <u>sets</u> of resistance values.

3) In the <u>E12 series</u>, there are <u>12</u> different values of resistance — these are 1.0, 1.2, 1.5, 1.8, 2.2, 2.7, 3.3, 3.9, 4.7, 5.6, 6.8 and 8.2 ohms (Ω). These can also be <u>multiplied</u> by a <u>power of 10</u>, e.g. $1.2 \, \Omega \times 10^3$ gives a $1200 \, \Omega$ resistor.

4) E12 resistors have a <u>tolerance</u> of $\pm 10\%$ — this means that the <u>actual resistance</u> is <u>within 10%</u> of the <u>value</u> stated on the resistor.

$\pm$ stands for plus or minus. E.g. $100 \pm 10\%$ means the value could be anywhere from 10% less than 100 to 10% more than 100.

EXAMPLE: A $100 \, \Omega$ E12 resistor has an actual resistance of $100 \, \Omega \pm 10\%$.
This means that the actual resistance is in the range of 10% below 100 to 10% above 100.
10% of 100 = 10. So the actual resistance is between $90 \, \Omega$ (100 − 10) and $110 \, \Omega$ (100 + 10).

If your circuit needs a $105 \, \Omega$ resistor, you would normally choose a $100 \, \Omega$ resistor as $105 \, \Omega$ falls <u>within</u> the $\pm 10\%$ <u>tolerance</u>. However, you may want to play it <u>safe</u> and choose a $120 \, \Omega$ resistor (which has resistance between 108 and $132 \, \Omega$) — this would <u>make certain</u> that the <u>current isn't too large</u>, as the <u>resistance</u> will <u>definitely be higher</u> than what you <u>need</u> (it'd be at least $108 \, \Omega$).

Colour of band	Number
Black	0
Brown	1
Red	2
Orange	3
Yellow	4
Green	5
Blue	6
Violet	7
Grey	8
White	9

5) <u>Coloured bands</u> are used to <u>identify</u> resistors. These are read from <u>left to right</u> (when the tolerance band is on the far right). The <u>colour</u> of each band <u>represents</u> a <u>number</u>.

6) There are <u>4</u> coloured bands on E12 resistors:

- <u>Bands 1 and 2</u> represent the <u>first two digits</u> of the value of <u>resistance</u>.

- <u>Band 3</u> represents the <u>number of zeros</u> that <u>come after</u> the <u>first two digits</u> in the resistance value.

- <u>Band 4</u> represents the <u>tolerance</u> — it's always <u>silver</u> (representing <u>10%</u>) for <u>E12 resistors</u>.

EXAMPLE: Identify a resistor with a yellow, violet, red and silver stripe.
Yellow = 4, violet = 7, red = 00 and silver = 10% tolerance.
The resistor = $4700 \, \Omega$ with a 10% tolerance.

Dual in Line IC Packages Have Two Rows of Pins

1) <u>Integrated circuits</u> (ICs) are <u>tiny</u> circuits that are <u>contained</u> within a <u>single component</u> (see p.26).

2) A <u>Dual in Line package</u> (<u>DIL</u>) is <u>one</u> of the ways that <u>ICs</u> are <u>packaged</u>:

- The circuit itself is placed inside a <u>rectangular protective casing</u> made of <u>plastic</u> or <u>ceramic</u>.
- <u>Metal legs</u> (<u>pins</u>) stick out of the casing in <u>two parallel rows</u>. The <u>number of pins varies</u> depending on the type of IC — commonly <u>between 8 and 40 pins</u>.

3) The pins <u>connect</u> the <u>IC</u> to the <u>rest of the circuit</u> — they're <u>pushed</u> through <u>drilled holes</u> in the PCB and are <u>soldered</u> on the <u>other side</u>. Alternatively, they can be <u>plugged into</u> an <u>IC socket</u> (a component that an IC slots into) so you don't have to <u>solder</u> the IC <u>directly</u> to the board — this means they can be <u>easily replaced</u>.

Casing
Pin

4) Each pin has a <u>number</u> — a <u>dot</u> or <u>notch</u> on the casing shows where the <u>1st pin</u> is, and they're numbered <u>anticlockwise</u> from that pin. This is useful as <u>most</u> pins have a <u>certain function</u>, e.g. some <u>only</u> connect to inputs.

Standard Components in Systems

PICs are a Specific Type of Microcontroller

1) Microcontrollers are programmable ICs that can do the job of multiple process blocks (see p.26) — a programmable intelligent computer (PIC) is a common type of microcontroller.

2) PICs are often sold as DILs (Dual in Line packages) and have a voltage rating of between 3 and 5.5 V.

3) Some PICs are one-time programmable (OTP) microcontrollers — they can only be programmed once.

4) However, most PICs have flash memory — they can be reprogrammed up to 100 000 times, so can be used again and again.

So when choosing a PIC, keep in mind that:
- Different PICs have different amounts of memory, types of memory and numbers of pins.
- PICs with more memory and more pins cost more to buy.

PICs with more pins can have more inputs and outputs.

Mechanical Components Come in Loads of Shapes and Sizes

Chain   Sprocket

Chains

1) Roller chains like the one on a bike are made up of metal links that are pinned together to form a loop. Chains transfer rotary motion from one sprocket to another.

2) Chains are sold by width and pitch (distance between the centres of the pins).

3) Chains only come in standard lengths — you may have to add or remove some links to make it the length you need.

Pin
Pitch   Link

Gears

1) Gears come in a variety of widths.

2) They can also have different numbers of teeth — this is important in a gear train to change the speed of rotation (see p.29).

3) Different types of gear will have different tooth shapes that interlock in different ways — for example, bevel gears have teeth angled at 45°.

Slower rotation   Faster rotation

Springs

1) There are 2 major types of springs that you can buy — compression springs (which resist compression) and extension springs (which, that's right, resist extension).

2) Springs can be found with different lengths, widths, and numbers of coils.

3) This can affect the spring rate of the spring, i.e. how easy it is to stretch or compress.

Extension
Compression

You've got to PIC a pocket or two — and no DILly-dallying...

I get it — these pages aren't thrilling. However, it's important that you know about some of the different components found in electronic and mechanical systems. Just be glad you don't need to know about them all...

Q1 Work out the resistance and tolerance of a resistor with bands that are coloured in the order: orange, white, brown, and silver. [2 marks]

Q2 What does DIL stand for? [1 mark]

Q3 Suggest one advantage of a PIC having flash memory. [1 mark]

Q4 Springs can be sold in varying lengths. Suggest two other features that springs can be sold by. [2 marks]

Cutting, Drilling and Soldering

Cutting and drilling processes are the bread and butter of most manufacturers — they're used all the time when working with metals, wood and plastics (see Section 5). Here are some examples that focus on systems...

Laser Cutters Cut Out System Components

1) Laser cutters use a laser to cut through materials.

2) The machine makes cuts by following a design that's loaded into it. This instructs it where to cut the material.

Computer-aided design (CAD) involves designing products using specialist computer software. Computer-aided manufacture (CAM) produces products from a design that's made using CAD (see p.4).

3) Laser cutters use designs made by CAD — they are CAM machines.

Like all CAM machines, laser cutters use computer numeric control (CNC). This involves using the precise coordinates set out in the design to control exactly where the laser cuts the material.

4) Laser cutters can only cut in 2D, so they have to be used on sheets of material — these can be sheets of plastic, wood, cardboard, fabric or some metals. Thicker sheets may need to be cut with more powerful industrial lasers.

5) Laser cutting has a huge range of uses — it can be used to cut out a variety of mechanical components (such as gears), and PCBs too.

6) It has high precision and accuracy — it's able to follow complex patterns, even on a small scale. This is helped by the tiny width of the laser beam.

7) However, a laser cutter requires a lot of power to run, so it can be expensive.

PCB Drills Cut Holes in PCBs

1) Printed circuit boards need to be drilled so that the pins or wires of each component can be pushed through the holes and soldered in place (see next page).

2) Printed circuit boards are often very small, so they need tiny holes. As a result, very thin drill bits are used — they can be as little as 0.2 mm wide.

3) Drilling PCBs can be a fiddly business — holes have to be drilled perfectly straight and in exactly the right spot. We therefore use a drill press (a small power drill that has a stand) rather than a hand-held drill to make it easier to drill the holes.

4) In industrial PCB production, CAM drills are often used, as they make the process even quicker and easier to do.

PCB with drilled holes — some have been filled with solder to attach components.

Laser-eyed rhinos charge lots — laser cutting isn't cheap you know...

There's some interesting stuff on this page, so hopefully it won't be a chore to learn. Make sure you know how computer-controlled laser cutting works, and that laser cutters and some PCB drills are CAM machines. Lovely.

Cutting, Drilling and Soldering

Soldering is used to <u>join metals</u> together — its <u>most common application</u> is in <u>mounting components</u> onto <u>PCBs</u>, so we'll be <u>focusing</u> on this here. Other applications include joining <u>plumbing pipes</u> and sealing <u>food cans</u>.

Soldering Provides an *Electrical Connection* in Circuits

1) Soldering uses <u>solder</u> to <u>attach</u> components to a circuit — it's used to <u>connect</u> a <u>component's metal pins</u> to either <u>wires</u> or the <u>copper tracks</u> on a PCB, and <u>hold</u> the components <u>in place</u>.

Some components don't have pins or wires. They are directly mounted to the surface of the PCB.

2) Solder is usually a <u>tin-based alloy</u>, and can be applied <u>manually</u> by <u>melting</u> it with a hot <u>soldering iron</u>.

3) It's <u>electrically conductive</u>, so it's used as an <u>electrical connection</u> between the <u>components</u> and the rest of the <u>circuit</u>.

Soldering iron

Soldering is also covered on p.71.

Badly soldered joints can have balls of solder rather than volcano shapes — these make poor electrical connections. Bad soldering can also involve using too much solder — this connects copper tracks that shouldn't be, which alters the circuit.

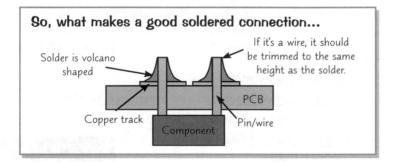

So, what makes a good soldered connection...

If it's a wire, it should be trimmed to the same height as the solder.

Solder is volcano shaped

Copper track
PCB
Component
Pin/wire

4) In <u>industry</u>, it would be <u>too slow</u> to <u>mass produce</u> PCB circuits using <u>manual soldering</u>. So <u>automated</u> soldering methods are used instead:

Flow Soldering

Direction of movement of the board
Solder wave
PCB

- This is when components are <u>placed</u> onto the PCB, which is then <u>passed over</u> a <u>pan of molten solder</u>. Sometimes components are <u>glued into place</u> before being passed over the molten solder.
- A <u>pump</u> produces an <u>upwelling</u>, or <u>'wave' of solder</u> at the right height to <u>just touch</u> the <u>base</u> of the board and <u>solder</u> each component <u>in place</u>.
- The whole process is <u>quick</u> and <u>cheap</u>, but a lot of <u>thought</u> has to go into <u>component layout</u> to make it work.

Flow soldering can also be called "wave soldering".

- It's generally only used for components with pins that <u>poke through</u> the PCB, e.g. DILs.

A solder volcano — the eighth wonder of the world...

Q1   Name three materials that laser cutters can cut.   [3 marks]

Q2   Suggest a downside of using laser cutters.   [1 mark]

Q3   Why do we need to drill holes into PCBs?   [1 mark]

Q4   a) State the two purposes of soldering components to a circuit board.   [2 marks]

       b) When manually soldering, suggest a piece of equipment you would use to melt the solder?   [1 mark]

PCB Production and Surface Treatments

It's time to <u>finish</u> PCBs off once and for all. It's not as bad as it sounds... just <u>etching</u>, <u>assembly</u> and <u>lacquering</u>.

Photo-Etching Removes any Copper we Don't Want as Tracks

<u>Photo-etching</u> converts a <u>blank PCB</u> (with a layer of copper over the <u>whole</u> board) into one with <u>custom copper tracks</u> that is <u>ready</u> for <u>drilling</u> and <u>soldering</u>. The process involves <u>UV light</u> and a <u>bucketful</u> of <u>chemicals</u> — it <u>removes all of the copper</u> from the board <u>except</u> for the <u>bits</u> that are needed to make the <u>tracks</u>.

1) The <u>layout</u> of the copper tracks is <u>designed</u> using <u>CAD</u> software — this is <u>printed</u> onto a sheet of <u>transparent material</u> (e.g. a plastic such as <u>acetate</u>) to form a "<u>PCB mask</u>".

The layers of a blank PCB:

Photosensitive layer

Copper layer — the electrically conductive layer that makes up the tracks.

Glass-reinforced plastic (see p.33) — this provides the base of the board and gives it strength.

2) The mask is placed onto the blank PCB on top of the <u>photosensitive layer</u> (see p.86).

3) The <u>board</u> is then <u>exposed</u> to UV for <u>3-10 minutes</u>. The <u>mask</u> is <u>opaque</u> in the <u>areas</u> of the board that will become <u>copper tracks</u>, so it <u>shields</u> these parts from the UV. The rest of the mask is <u>transparent</u>, so UV can <u>travel</u> through to the <u>photosensitive layer</u>.

4) The board is then exposed to two different <u>chemicals</u> separately:
   - <u>Developer</u> — this <u>dissolves</u> parts of the <u>photosensitive layer</u> that were <u>exposed to UV</u>.
   - <u>Etching chemical</u> — this <u>dissolves</u> any <u>exposed copper</u> (that was just revealed by the developer).

5) Copper <u>still covered</u> by a photosensitive layer <u>isn't removed</u> from the board — these are the areas that'll become <u>copper tracks</u>.

6) Finally, any remaining <u>photosensitive layer</u> is <u>stripped off</u>, exposing the bare copper tracks. Now the board is ready for <u>components</u> to be <u>added</u>...

PCBs can be Assembled Manually or Using a Machine

1) <u>Pick and place assembly</u> is the process of <u>placing components</u> onto a <u>PCB</u> in the <u>correct position</u> and <u>orientation</u> — so they're ready to be soldered onto the board.

2) This can be done using a <u>pick and place CAM machine</u> or <u>by hand</u>:

Pick and Place Machines

1) Pick and place machines use <u>suction cups</u> to <u>pick up</u> components and <u>move them</u> to their <u>correct location</u>. The cups <u>rotate</u> so the components are the <u>right way round</u> before being placed onto the board.

2) Machines need to be <u>programmed</u> with the <u>rotation angle</u> and the <u>location</u> of <u>every component</u> on the PCB. This can be a <u>lengthy</u> process.

3) These machines are much <u>quicker</u>, more <u>precise</u> and more <u>consistent</u> than assembling PCBs manually. In <u>mass production</u>, this <u>makes up</u> for the <u>high cost</u> of the <u>machines</u>.

4) Machines are normally only used for <u>surface-mounted</u> components.

Manual Assembly

1) Placing the components by hand can be a <u>fiddly</u> business, especially when components are <u>small</u> — <u>tweezers</u> are often used to <u>increase precision</u> of placement.

2) This is generally used for components with <u>pins</u> that need to be <u>pushed through holes</u>, as machines <u>can't do this</u> very <u>easily</u>.

3) Manual assembly often takes <u>longer</u> than machine assembly (as <u>humans</u> need <u>regular breaks</u>) and is likely to include <u>more mistakes</u> (due to <u>human error</u>).

PCB Production and Surface Treatments

PCB Lacquering Protects Components from Corrosion

1) A PCB lacquer is a surface treatment for PCBs — it's a thin polymer film that provides a protective barrier against moisture, chemicals, large temperature changes and dust.

2) This is important as electronic components often need to cope with harsh environments — it helps to decrease corrosion (see p.73) and increase the durability of the board.

3) PCB lacquers can be sprayed on, painted on or applied by dipping the whole board into the lacquer.

4) Many lacquers are fluorescent — they glow in UV light. This is used in quality control (see p.40), as it makes it easy to see where a coating has been applied and more importantly, if any places have been missed.

Lubricants Reduce Friction Between Moving Parts

1) Friction is the force that stops two things from sliding past each other. They need to be touching for friction to act. For example, when a bike brakes, the brake pads come into contact with the wheels and the friction between them eventually stops the wheels from turning.

2) Friction can cause problems in a lot of mechanical systems:

   • It can reduce the efficiency of the system — moving parts that are linked or interlock are slowed down, causing kinetic energy (movement) to be partially converted to waste heat energy.

   • It can lead to interlocking parts being worn down, increasing the chance of them becoming damaged and reducing their lifespan.

In an efficient system, only a small amount of the energy that's released goes to waste (e.g. waste heat energy) — the rest is useful.

3) Mechanical systems are therefore often lubricated to reduce friction — a slippery material called a lubricant is applied between the two surfaces to allow them to slip past each other easily.

4) Importantly though, lubricants don't get in the way of the mechanism — for example, gears will still interlock and work as normal, but just more efficiently and with less wear and tear.

There are loads of types of lubricant for all kinds of applications:

   • Graphite (the material used for pencil leads) is used to lubricate locks.

   • Oil is used on bike chains and in vehicle engines.

   • Silicone or grease is often used to lubricate door hinges and stop them from squeaking.

Crikey, I wanted my back itched — not etched...

Yay! Finally finished the section — it's pretty hard stuff, so well done for getting through it... Time for a tea break.

Q1 a) Describe how a PCB mask is made. [1 mark]

   b) Describe how the mask is used in the photo-etching process of PCBs. [2 marks]

Q2 Karthik is looking to assemble 10 identical printed circuit boards on a tight budget. He's using components that need to be pushed through holes. Describe why he should assemble the PCBs by hand rather than using a machine. [2 marks]

Q3 Give two things that a PCB lacquer protects against. [2 marks]

Q4 Give one example of a lubricant. [1 mark]

Q5 Give two examples of mechanical systems that may need lubrication. [2 marks]

Looking at the Work of Designers

You have to complete a design and make task as part of your non-exam assessment (NEA) (see p.1), but there's also a section of the exam dedicated to designing and making principles. So now seems like a good time to get clued up on all things design — there's lots to consider. First up — looking at the work of existing designers...

Looking at the Work of Others Can Give You Ideas

1) Being influenced by good designers is a great way to get ideas.
   You can't just copy their stuff though — that's illegal.

2) You need to know about the work of at least two of the designers listed below and on the next page.

3) They've all developed styles that have influenced other designers. Investigating, analysing and evaluating the work of some of these designers could influence your designs too.

4) Here's a bit of information on each designer to get you started...

HARRY BECK — (1902-1974) — redesigned the London Underground map in the 1930s. Its simplified layout made it a huge success — maps of many other transport networks now use Beck's style.

VIVIENNE WESTWOOD — (1941-Present) — her iconic clothing became very popular during the punk rock movement in the 1970s. She has since become a world famous fashion designer. Her designs often take inspiration from traditional British clothing and historical paintings.

MARCEL BREUER — (1902-1981) — a modernist architect and furniture designer. Some of his best known work includes furniture made from tubular steel and buildings which look like they've been sculpted from concrete.

> Modernist design is simple with no ornamentation — it focuses on the function of a product rather than its appearance. Modernist buildings were typically made from concrete, steel and glass.

MARY QUANT — (1934-Present) — a fashion designer who popularised the mini skirt, hot pants and PVC in the sixties. Her clothing often featured white collars, simple shapes and bold colours.

NORMAN FOSTER — (1935-Present) — an architect whose buildings are often constructed from glass and steel. Foster's work includes famous landmarks such as Wembley Stadium, the Millennium Bridge and 'the gherkin' in London.

WILLIAM MORRIS — (1834-1896) — a designer who is best known for his wallpaper, furniture and furnishings. His designs were often based on patterns found in nature.

ALEXANDER MCQUEEN — (1969-2010) — an influential fashion designer known for his theatrical, well tailored clothing and dramatic catwalk presentations displaying his collections.

ALDO ROSSI — (1931-1997) — an architect who also published work on architectural theory. He designed a number of products too, including some for Alessi (see the next page).

PHILIPPE STARCK — (1949-Present) — an architect and product designer. He has designed many products including furniture, kitchenware and vehicles. One of his best known products is his lemon squeezer for Alessi.

> Art Nouveau designs are flowing and curvy. They often use floral or insect motifs.

LOUIS COMFORT TIFFANY — (1848-1933) — an Art Nouveau jewellery and glassware designer. He is best known for his work with leaded glass, which he used to produce windows and lampshades.

> The Art Deco design style involves bold colours, geometric, zigzag and stepped shapes, bold sweeping curves and sunburst motifs.

RAYMOND TEMPLIER — (1891-1968) — designed Art Deco jewellery. His geometric pieces were different to the traditional styles of the time.

COCO CHANEL — (1883-1971) — a fashion designer known for introducing practical, casual-chic clothing for women, who had traditionally worn corsets and long skirts.

Looking at the Work of Designers

CHARLES RENNIE MACKINTOSH — (1868-1928) — an architect and designer. His work gained most recognition after his death, but his unique style is shown in a number of buildings of his design in Scotland, including the Glasgow School of Art.

GERRIT RIETVELD — (1888-1964) — an architect and designer involved in De Stijl (Dutch modernism). De Stijl designs were basic — they used simple shapes, vertical and horizontal lines and primary colours.

SIR ALEC ISSIGONIS — (1906-1988) — a car designer who designed the Mini in the 1950s and the Morris Minor in the 1940s.

ETTORE SOTTSASS — (1917-2007) — an architect and product designer. He was the founder of the Memphis movement. Memphis designs used many bright, contrasting colours as shown by the Carlton room divider, designed by Sottsass. He also designed Alessi products.

The Memphis movement was at the height of postmodernism. Postmodernist designers were sick of modernist design and thought that style should be the starting point of a product.

You need to investigate the work of at least two of these companies too...

Braun	1)	Braun is an electrical appliance company based in Germany.
	2)	Dieter Rams, who oversaw design at Braun for over 30 years, worked on the basis of 'less, but better' — his designs were simple, with no unnecessary details. This made Braun's products easy to use and gave them an industrial look.
Dyson™	1)	Sir James Dyson invented the first bagless vacuum cleaner in the 1980s. Before this, vacuum cleaners lost suction over time and contained bags to collect dust which needed replacing.
	2)	Dyson's design made use of cyclones and a dust bin (a plastic container where dust collects). This gave greater, consistent suction and there was no need to keep buying bags.
	3)	Dyson™ have since created a variety of vacuum cleaners designed with a focus on being efficient and innovative. The company also produces a range of other products.
Apple®	1)	Apple® products are instantly recognisable due to their smooth, rounded edges and simple and sleek style. Their products are very intuitive, making them user friendly.
	2)	Sir Jonathan Ive, who is responsible for design at Apple®, used Dieter Rams' work as inspiration. This 'less, but better' attitude is shown in the sleek design of Apple® devices.
Alessi	1)	Alberto Alessi took over the running of his family's kitchenware company in the 1970s. He didn't design any products himself — he employed designers and architects to come up with fun and creative designs which were then manufactured by his company.
	2)	A key part of his idea was to mass-produce products but keep them as stylish and original as possible. The designs were always distinctive and often colourful. A well known example is Philippe Starck's lemon squeezer.
Gap	1)	Gap is an American clothing company which sells casual, everyday wear (e.g. jeans, t-shirts, etc.).
	2)	The company started out in 1969 as single shop, initially just selling jeans alongside records (to attract young people into the shop). Its popularity led to the company quickly increasing its number of shops and range of clothing. The company now has thousands of stores worldwide.
ZARA	1)	Zara is a Spanish clothing company which sells up-to-date, fashionable clothes at affordable prices.
	2)	The company constantly responds to evolving fashion trends and customer feedback. This means their clothing ranges change frequently, keeping their customers coming back to see what's new.
Primark®	1)	An Irish clothing company which opened its first store in Dublin (under the name of Penneys) in 1969. Primark® now has stores across the UK, in Europe and the USA.
	2)	Primark® aims to provide current trends in fashion at a low cost to its customers.
Under Armour®	1)	Under Armour® is an American sportswear company. It was founded by Kevin Plank, an American footballer. He created a shirt that wicked sweat away from the skin after becoming fed up with his sportswear absorbing large amounts of sweat during exercise.
	2)	His shirts were very popular with other football players, leading to the company expanding. Under Armour® now sell a wide range of sportswear and fitness devices.

I'm looking forward to the postrevision movement...

Q1 Give an example of a designer who designed household items. [1 mark]

Understanding User Needs

Different people have <u>different needs</u> (they also have <u>different values</u> — see p.11).
These needs must be <u>taken into account</u> when <u>designing</u> a <u>product</u> for a particular <u>group of people</u>...

Ergonomics Means Making the Product Fit the User

Designing a product so that it's <u>comfortable</u> and <u>easy to use</u> is known as <u>ergonomics</u> (or <u>human factors</u>).

1) Products need to be designed so that their <u>size</u> and <u>proportions</u> fit the needs of the user. For example, the buttons on a calculator need to be big enough for the user to press them individually. Textile products need to <u>vary in size</u> (e.g. clothes) or be <u>adjustable</u> (e.g. rucksacks) to fit the user.

2) Ergonomic design also helps to ensure that <u>using</u> the product won't cause <u>health problems</u>. For example, an ergonomically designed chair should prevent you suffering backache from using the chair regularly.

3) Ergonomics needs to take the <u>target market</u> into account, e.g. a chair for a five-year-old needs to be a different size from a chair for a fifteen-year-old (obviously).

4) Designers use <u>body measurement data</u> (also called <u>anthropometric data</u> — see below) to make sure the product is the <u>right size</u> and <u>shape</u>.

Many chair designs achieve this by allowing you to have your feet on the floor with your knees at a right angle and your back supported.

Anthropometric Data are Measurements of Humans

1) <u>Measurements</u> of <u>human body parts</u> are called <u>anthropometric data</u>. This includes things like:

- Upper arm length
- Head circumference
- Height (stature)
- Chest depth
- Shoulder height
- Hand width
- Knee height
- Little toe length
- etc...

2) These <u>measurements</u> are collected from a <u>wide range</u> of people with <u>different body sizes</u>.

3) To make <u>products</u> of the <u>right size</u>, designers need to know the <u>likely body measurements</u> of the <u>intended users</u>...

1) First, they work out <u>what measurements</u> they need. For example, if their product is a novelty mask it doesn't matter how long the users' legs are — they only need to know about their heads.

An adult's novelty <u>mask</u> has to be <u>big enough</u> to cover people's faces, but there are lots of fiddly details to consider...

The eye holes have to be the right distance apart.

The raised part for the nose has to be big enough.

If the mask was for children, a different (smaller) set of anthropometric data would be used.

2) Next, they <u>find out</u> what these measurements are for the <u>typical user</u> of the product — these measurements will have been <u>collected</u> by <u>sampling</u> lots of people from the <u>target group</u>.

3) Then, they design the product to fit a <u>range of users</u> within the <u>target group</u> (e.g. adults for the novelty mask). There's more on how this is done on the next page...

Understanding User Needs

Products are Often Designed to Fit 90% of the Target Market

1) Manufacturers try to design products so that they'll be suitable for <u>most people</u> in the <u>target market</u> — they often aim to make the product fit <u>okay</u> for <u>90%</u> of the target users.

2) For example, a manufacturer designing an office <u>desk</u> might look at data for the <u>heights</u> of a sample of adults aged between 20 and 65:

90% of people are taller than 1.64 m but shorter than 1.86 m.

Number of people

1.64 m
5th percentile

50th percentile

1.86 m
95th percentile

Height

The 50th percentile is the average — 50% of people are shorter and 50% of people are taller.

1) Manufacturers often try to make 'one size fits all' products, but actually, it's 'one size fits 90%' — they use the <u>5th and 95th percentiles</u> as cut-off points:

5th percentile: this means that <u>5%</u> of people are <u>shorter</u> than this value.
95th percentile: this means that <u>95%</u> of people are <u>shorter</u> than this value.

2) Here, this means the <u>bottom 5%</u> (very short people) and the <u>top 5%</u> (very tall people) <u>aren't catered for</u>. A one-size product wouldn't suit either extreme.

3) These people would probably have to use a specialist supplier or have products <u>custom-made</u> for them.

Products Need to be Accessible for Disabled Users

Lots of products are specifically designed to help people with <u>disabilities</u>.

1) Some packaging (e.g. for medication) has <u>Braille</u> labelling to give blind people information.

2) Control buttons can be made <u>brightly coloured</u> and <u>extra large</u>, so they're easy to find and press. For example, telephones, TV remotes and calculators can be made with big buttons.

3) Products such as smoke alarms can be designed with <u>visible</u> signals as well as audible ones so that deaf people can be alerted to fires.

4) Designers also have to think about <u>wheelchair users</u>. For example, trains and buses need to be designed to have wheelchair access.

Designers Need to Think About Age Groups

People in different <u>age groups</u> have different <u>physical limitations</u>. For example:

1) Small children and elderly people may not be able to manipulate <u>small parts</u> and may have difficulty undoing <u>fastenings</u> and opening <u>packaging</u>.

2) Elderly and infirm people might have difficulty <u>holding</u> and <u>using</u> products. Designers can think about putting large, easy-to-grip handles on, say, cutlery.

<u>Older people</u> may also have different <u>requirements</u> from <u>textile products</u>, e.g. they may prefer shoes with <u>lower heels</u> and <u>wider fittings</u> for comfort.

Sam was having trouble holding the fish — the handle was too small.

Make sure you're comfortable with these pages...

Remember, designers need to think carefully about who a product is aimed at and make sure it's suitable for them.

Q1 a) Suggest three types of anthropometric data that might be needed when designing a football shirt. [3 marks]

b) Why might a claim that a football shirt is 'one size fits all' not be true? [1 mark]

Q2 Suggest how a can opener could be made suitable for infirm elderly people. [1 mark]

Design Briefs and Specifications

The process of designing and making something is called the <u>design process</u> (gosh).

Designing Starts with the Design Brief

So, someone gets an idea for a <u>new product</u>.
They decide to <u>employ a designer</u> to work on the idea.

1) The person who hires the designer is called the <u>client</u>.

2) The <u>client</u> gives the designer a <u>design brief</u>...

3) The design brief is a <u>starting point</u> for the development of the product. It should include:

- what <u>kind</u> of product is needed or wanted (and <u>why</u>)
- how the product will be <u>used</u>
- <u>who</u> the product is <u>for</u> (the <u>target market</u>)

4) The designer needs to pick out all the <u>important features</u> of the design brief.

5) One way of doing this is by drawing a <u>spider diagram</u>. It's a quick way to help analyse the problem and to help the designer decide what they need to <u>research</u>.

DESIGN BRIEF FOR BACKSCRATCHER/TURNIP HOLDER

No currently commercially available backscratcher has an in-built capacity for turnip storage. We want you to design a product to meet this need for those people having itchy backs and modest turnip storage requirements (up to 4 turnips).

A design doesn't have to be based on the needs or wants of a <u>client</u> — a product idea can be based on <u>your</u> wants, needs and interests too.

Research Helps the Designer Get Ideas

The point of doing research is to:

1) check that people will actually <u>want the product</u>.

2) find out what people in the target market <u>like or dislike</u> about similar <u>existing products</u>.

3) find out what <u>materials</u>, standard <u>components</u> and <u>techniques</u> would be suitable for the product, and how they will affect the manufacturing and selling <u>costs</u>.

There are <u>different types</u> of research:

Market Research

<u>Questionnaires</u> or <u>interviews</u> can be used to find out people's likes/dislikes and so on. This helps the designer understand what the <u>target group</u> wants from a product.

There's loads more on carrying out research on pages 100-103.

Product Analysis

Designers can <u>examine</u> a current product or <u>disassemble it</u> (take it <u>apart</u>) to:

- figure out how it's <u>made</u> (e.g. materials, components and processes) and how it <u>works</u>.
- check for <u>good</u> and <u>bad</u> features, including how ergonomic the design is (see page 102).
- <u>collect data</u> on factors that could affect the product's <u>performance</u>, e.g. its <u>size</u>, <u>weight</u> etc. Several products can then be <u>compared</u> at once using <u>tables</u> or by plotting the data on <u>graphs</u>.
- test how it <u>feels</u>, <u>looks</u>, <u>smells</u> or <u>tastes</u> (sensory analysis).

Then they should have a good idea of what <u>materials</u> and <u>techniques</u> they need to research for the product.

I've designed your briefs, now on to the socks...

Designers don't just go straight from the design brief to sketching ideas. They first need to check out other similar products already on sale and get some information about the target market for the new product. First things first.

Design Briefs and Specifications

There's loads more on how to actually carry out research on the next few pages.
But before we get onto that, let's take a look at what designers do with their research findings...

Research is Used to Draw Conclusions

Once they've done some market research and product analysis, designers should have loads of information. They use the information to help with their design. They should:

1) Summarise what they've found out — pick out the most important and useful findings.
E.g. carrots are just as popular as turnips. This might involve interpreting data and graphs.

2) Explain what impact each finding will have on their design.
E.g. the design should be able to hold carrots as well as turnips.

The Findings Might Mean the Design Brief is Reconsidered

1) Once conclusions have been drawn from the research, the designer might want to look again at the design brief with the client — remember, the design brief is just the starting point.

2) For example, the research might show that most people want the turnip holder/backscratcher to be collapsible for easy transportation or store a minimum of 8 turnips. As a result, the designer may have to consider altering the design brief, with any decisions made being discussed with the client.

The Design Specification is a List of Conditions to Meet

1) Once the design brief is agreed, a design specification can be made.

2) The design specification gives certain conditions that the product must meet. These conditions (often called design criteria) should take account of the research findings.

3) It's best to write a specification as bullet points. Each point should be explained and show how the research helped the designer to decide on that point. The specification should cover the following things:

Tally chart showing the minimum length of backscratcher that people needed to scratch the middle of their back.

Length (mm)	No. of people
150	\
200	\\\\
250	ⵌ \\
300	ⵌ ⵌ
350	ⵌ \
400	\\\\
450	

- Size — how big it is
- Aesthetics — how it looks
- Consumer — who will buy it
- Function — what it will do
- Quality — e.g. the required finish
- Cost — the price range
- Materials — what it is made of
- Safety — how to make sure it's safe
- Environment — the impact on the world
- Sustainability — its future impact

See page 103 for more on why environmental issues need to be considered in your design specification.

Example:

SIZE
- It should weigh 300 g or less, as most people were comfortable with this.
- The minimum length will be 400 mm, as that was long enough for everyone to reach the middle of their back.

AESTHETICS
- It should be multicoloured because most people said they would like the product to be very colourful.
- The handle should be easy to grip but not feel rough, because some people were concerned that a rough surface might catch on their clothes.

Design specs — not a type of fancy glasses...

Q1	What is the name given to a person who hires a designer?	[1 mark]
Q2	What information should a design brief include?	[3 marks]
Q3	Give one example of a type of market research.	[1 mark]
Q4	List two things you can find out about a product during product analysis.	[2 marks]
Q5	What is a design specification?	[1 mark]

Market Research

As we said on the previous two pages, <u>market research</u> is important for <u>designing</u> a product that people will <u>want to buy</u>. Now it's time to go into more <u>detail</u> about how exactly market research is <u>carried out</u>...

Identify your Target Group

1) Even the very best products won't be everyone's cup of tea — some people will <u>like</u> them, some <u>won't</u>.

2) You might be given a <u>target market</u> in the design brief, or have a specific <u>target group</u> you want to design for. If not, you need to work out which people are <u>most likely</u> to buy your product — this should be your target group. Ask members of your target group <u>what they want</u> the product to be like.

3) You can group people by things like <u>age</u>, <u>gender</u>, <u>job</u>, <u>hobbies</u>, <u>lifestyle</u>, <u>amount of money</u> <u>they earn</u>, or anything else — it'll probably be a combination of a few of these things.

> For example, if you're trying to sell <u>wide-legged, elasticated-waist, linen trousers</u>, you may decide to target them at <u>middle-aged women</u> who are buying their <u>summer clothes</u>.

> But if you've designed some fetching <u>stripy knee-length socks</u>, you'd be better off targeting them at <u>teenage girls</u> in winter.

Think Carefully About What You Want to Find Out

Once you're clear on exactly who your target group are, you need to decide what to <u>ask</u> them. You could find out:

1) Some information about the <u>person</u> answering your questions. This could help you make sure they're within your <u>target group</u>, or give you <u>extra information</u>.
   - Are they male or female? (probably best to judge for yourself rather than asking...)
   - What age bracket are they in? (11-15, 16-20, 21-25 etc.)
   - What job or hobbies do they have?

They never stock clothes in my size...

2) Do they already <u>buy</u> the kind of product you're thinking of developing?

3) Do they like a particular <u>style</u> or <u>colour</u>?

4) <u>How much</u> would they be prepared to pay for this kind of product? This could affect your <u>budget</u> — the lower the selling price, the lower the manufacturing costs will need to be to make a profit.

5) <u>Where</u> would they expect to buy it from? Again, this could affect costs. E.g. if the majority of your target group buy cheap items of furniture from IKEA®, you should also design a low-cost item.

6) Is there something they <u>would like</u> from the product that existing products <u>don't have</u>?

> At a <u>later stage</u> in the design process, when you've generated <u>ideas</u>, you may decide to ask the target group <u>more questions</u>.
> E.g. <u>will they buy your version of the product?</u> Explain the advantage of your product over existing ones — would that be enough to tempt them to buy your version?

Questionnaires are Forms for People to Fill in

Now, how to phrase those all-important <u>questions</u>. There are <u>two basic types</u> of question:

1) <u>CLOSED QUESTIONS</u> — these have a <u>limited</u> number of possible answers, e.g. '<u>do you ever use a bag?</u>' can only be answered 'YES' or 'NO'. Analysing is easy, and you can show clear results at the end. Closed questions include <u>multiple-choice questions</u> — these give a <u>choice</u> of answers. Sometimes the person answering can pick more than one. E.g.

Weirdo...

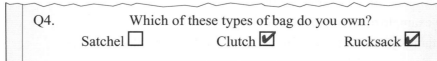

Q4.	Which of these types of bag do you own?	
Satchel ☐	Clutch ☑	Rucksack ☑

Market Research

2) <u>OPEN QUESTIONS</u> — these have <u>no set answer</u>, e.g. <u>what's your favourite type of bag, and why?</u> They give people a chance to provide details and opinions. This type of questioning takes more time and it's harder to draw conclusions from the results. But you could gain valuable information.

Interviews are Face-to-Face Conversations

1) For interviews, you can <u>start off</u> by asking the same sort of questions as in questionnaires — but then take the chance to ask <u>follow-up</u> questions, based on the answers you get.

2) Get your interviewees to give you <u>extra information</u> to explain their answer — this might help you get more <u>ideas</u> for your product. E.g. if their favourite type of wood is oak, ask them <u>why</u> they like it.

3) Interviews can give you more <u>detailed</u> information than questionnaires — you can have short <u>conversations</u> with people you're aiming to sell to. Just make sure you <u>stick to the point</u>.

4) A problem with interviews is that it's often more <u>difficult</u> to <u>analyse</u> the results than with questionnaires — it's harder to <u>compare opinions</u> when the topics covered vary slightly from person to person.

5) It can be useful to <u>record</u> the interview while it is taking place — audio recordings mean that the interview can be <u>analysed at a later date</u>, and it means that the person asking questions doesn't need to <u>stop the flow of conversation</u> to take notes. If a visual recording is also taken, then the person's <u>body language</u> can also be <u>analysed</u> to help interpret their answers.

Focus Groups Provide a Deeper Level of Feedback

1) Holding a <u>focus group</u> is a bit like carrying out a <u>group interview</u> — the group is asked for their <u>opinions</u> on either the proposed product <u>in general</u>, or one <u>particular aspect</u> of it such as the <u>packaging</u>. Like interviews, focus groups are a great way of getting more in-depth <u>feedback</u> from your <u>target group</u>.

2) The group should be made up of people who <u>don't already know each other</u>. This is important, as people are often more likely to <u>say how they feel</u> with <u>strangers</u>, whereas people they know may <u>influence</u> their <u>feedback</u>.

3) The aim of the focus group is to <u>generate a guided discussion</u> — where members can <u>freely</u> give their opinions to questions that have been <u>prepared in advance</u>.

4) Each group should have a fairly <u>small number</u> of <u>people</u> (e.g. 6-10) and be asked a set of <u>open-ended questions</u>. It's also a good idea to hold <u>more than one</u> focus group so that <u>results can be compared</u>.

ICT can be used to Research, Analyse and Present Information

1) You could write a <u>questionnaire</u> using <u>word processing</u> software — so it'll be <u>neat</u> and easy to read. This could be posted on a <u>website</u> that your <u>target group</u> is likely to use, so they could fill it in <u>online</u>.

2) You could use the <u>internet</u> to look up <u>sales figures</u> and collect them in a <u>spreadsheet</u>.

3) You could use <u>spreadsheets</u> to <u>organise and sort data</u> (e.g. your questionnaire results). Spreadsheets also allow you to <u>analyse</u> your data and <u>present</u> it using <u>charts</u> and <u>graphs</u>.

 EXAMPLE:

> Jordan is conducting market research for her new flip-flops, and has received 35 responses to her questionnaire. 20% of the 35 people that responded said their favourite flip-flop colour was orange. How many people said their favourite flip-flop colour was orange?
>
> 1% of people = 35 ÷ 100 = 0.35
>
> so 20% of people = 20 × 0.35 = <u>7 people</u>
>
> In a spreadsheet, you can use a formula to do this calculation.

My target group is tall, dark and handsome...

It might seem like a lengthy process, but doing your research properly is key to creating a product that is going to sell well. You should think carefully about who your target group is and what features are most important to them.

Q1 Describe a potential drawback of asking open questions on a questionnaire. [1 mark]

Q2 Suggest a way in which a spreadsheet can be used to help carry out market research. [1 mark]

Product Analysis

Next up, it's <u>product analysis</u> — this is where you <u>research</u> and <u>analyse</u> some <u>existing products</u> to give you a few <u>ideas</u> on what to use in your own designs. It will help you to write the <u>design specification</u> too...

There's Lots to Consider When Analysing a Product

1) As mentioned on p.98, <u>product analysis</u> can be done by looking at the <u>outside</u> of a product or <u>taking it apart</u>. There are <u>several</u> things to consider:

Function

1) Function is what the product is <u>intended</u> to do — its job.
2) <u>Disassembling</u> (taking apart) a product can help you find out <u>how it works</u>.
3) Make careful notes as you disassemble something, and record what <u>component parts</u> have been used and how it's <u>structured</u>, using <u>sketches</u> or <u>photos</u>.

Form

1) This is the <u>shape</u> and <u>look</u> of the thing, e.g. colour, texture and decoration.
2) A product could be <u>old-fashioned</u> or <u>modern-looking</u>. It could have <u>flowing curves</u> or it might be very <u>angular</u> with lots of corners. The general look of a product is also known as its <u>aesthetics</u>.

Ergonomics is also covered on page 96.

Ergonomics

1) Ergonomics is about <u>designing</u> products so that their <u>size</u> and <u>proportions</u> fit the users' needs.
2) For example, a <u>hand-held</u> product such as a touchscreen phone needs to <u>fit well</u> in one hand, and many are designed so that <u>all parts</u> of the <u>screen</u> can be <u>reached</u> with your thumb.
3) Designers use <u>body measurement data</u> (<u>anthropometrics</u>) to make sure the product is the <u>right size</u> and <u>shape</u>.

Competition and cost

1) You need to consider <u>value for money</u>. For example, if you're looking at a <u>hairdryer</u>, find out whether it's cheaper or more expensive than <u>similar</u> hairdryers.
2) You'll also need to look at <u>how it performs</u> compared to these other hairdryers.

Sustainability

<u>How much</u> does <u>making</u> or <u>using</u> the product harm the <u>environment</u>? E.g. most cars <u>emit carbon dioxide</u> (which causes global warming) and <u>various other pollutants</u>.

There's more about sustainability on the next page.

Materials

Product analysis should include looking at <u>what materials</u> have been used, and <u>why those materials</u> were chosen.

Manufacture

1) Consider all the <u>processes</u> that have been used to make the product.
2) This includes things like which <u>techniques</u> were used to <u>shape parts</u> of a product, e.g. the pieces of a garment.
3) Don't forget to check if any parts have been assembled separately and plonked into the product later — the term for that is <u>sub-assembly</u>.

2) It's important that you <u>record</u> the findings of your analysis — <u>make notes</u> so you can remember what you thought, and <u>compare</u> it to other products you look at. You can use a spreadsheet to record data.
3) Once you've considered all these aspects of the product, you should make an <u>evaluation</u> — a <u>judgement</u> about what does or doesn't work about the product, and how <u>effective</u> the different features are.

Analysing products — an excuse to play with things all day...

You probably think this seems a lot to think about before starting on the design of a product. But it's all worth doing. There's no point coming up with a new product if it's no better than an existing one...

Product Analysis

Which Materials are Used Matters

You need to consider the environmental impacts of materials when doing product analysis:

1) Some materials are toxic, e.g. some paints, varnishes and dyes.

2) Many materials are made from finite resources. For example, most plastics and synthetic fibres are made from fossil fuels (see p.12), which will eventually run out.

3) Products that use recycled materials, e.g. recycled metals and plastics, are more environmentally friendly.

4) It's also better to use sustainable materials, e.g. wood, paper and cotton. Softwoods (which can be regrown in a person's lifetime) are a better choice than hardwoods (which take ages to grow). Wood and paper from ethical sources (e.g. sustainably managed forests) won't contribute to deforestation (see p.47).

5) Many products are thrown away — it's good if these products are made from biodegradable materials (materials that will rot away naturally) or recyclable materials. For example, waste wood is biodegradable and can also be recycled into manufactured boards.

Processes Have Environmental and Social Impacts Too

You should also think about the processes used to make products:

1) Does the manufacturing process cause pollution? For example, how much waste material will be produced and how will it be disposed of?

2) Does the process use a lot of energy? If manufacturers wanted to be extremely responsible, they'd try to use renewable energy sources like wind power or hydroelectricity.

3) Was the product made under good working conditions? For example, does the manufacturer pay workers fairly, give them protective clothing, etc.

Fairtrade Makes Sure Producers get a Fair Price

1) A Fairtrade product, or product containing Fairtrade materials (e.g. Fairtrade gold or Fairtrade cotton), can have a smaller social and environmental impact.

2) Fairtrade producers must meet particular standards. This gives benefits to workers, the environment and the producers themselves:

Social standards provide:
- Fair terms of trade (higher prices) for the produce of small-scale farmers and workers, allowing them to earn a better income so they can afford to feed their families.
- Safer working conditions for farmers and workers.
- Fairtrade premiums — money for farmers and workers to help their community in an area they wish to improve, e.g. schools, roads, healthcare etc.

Environmental standards aim to reduce environmental impact in ways such as:
- reducing the use of pesticides
- reducing greenhouse gas emissions
- protecting biodiversity.

3) For these reasons, designers may deliberately choose to use Fairtrade materials in their product over those with fewer social and environmental benefits.

I'll swap you a ginger nut for a brownie — that's a fair trade, right...

Q1 Give three things that should be considered when analysing a product. [3 marks]

Q2 What is the term for assembling parts separately and then adding them to the product later? [1 mark]

Design Strategies

You can use <u>design strategies</u> to come up with <u>initial design ideas</u> without you <u>getting stuck</u> on a <u>bad one</u>...

There are Several Different Design Strategies

Designing is a really <u>complex</u> process and there are several different ways of doing it:

> **Systems Approach**
> This means <u>breaking down</u> the <u>design process</u> into a number of <u>different stages</u> and doing each in turn —
> e.g. writing the <u>design specification</u>, coming up with <u>ideas</u>, <u>developing</u> your ideas, <u>research analysis</u>, etc.
> This is a very <u>orderly</u> and <u>reliable</u> method of designing.

> **User-Centred Design**
> In this strategy, the <u>wants</u> and <u>needs</u> of the <u>user</u> are <u>prioritised</u> — their <u>thoughts</u>
> are given lots of <u>attention</u> at <u>each and every stage</u> of the design process.

> **Iterative Design**
> This is the design strategy that you'll use to <u>make</u> your <u>prototype</u>
> during your <u>NEA</u> (non-exam assessment). It is centred around a
> <u>constant process</u> of <u>evaluation</u> and <u>improvement</u> — see below.

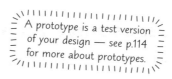
A prototype is a test version of your design — see p.114 for more about prototypes.

Iterative Design is a Circular Process

1) The iterative design process <u>doesn't stop</u> once you've come up with your <u>first good design</u> and made a
   <u>prototype</u>. It still needs to be <u>tested</u>, <u>evaluated</u> and <u>improved</u> based on the <u>results of testing</u>.

2) The whole process is then <u>repeated</u> for the <u>new</u> and <u>improved prototype</u>, so it can get pretty <u>circular</u>...

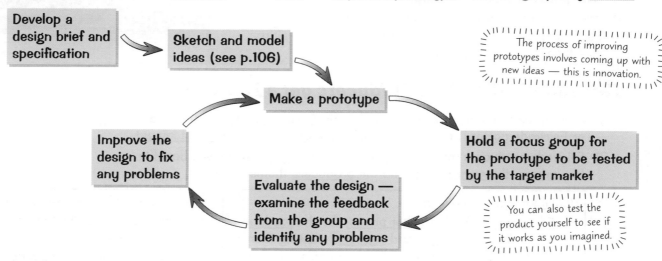

The process of improving prototypes involves coming up with new ideas — this is innovation.

You can also test the product yourself to see if it works as you imagined.

3) This process is <u>repeated</u> until all the <u>problems</u> have been <u>identified</u> and then <u>fixed</u>.
   Then the product is ready for production.

You can Consult Other People During the Design Process

1) <u>Collaboration</u> is all about working with other people. It allows you to gain
   <u>specialist knowledge</u> from <u>across</u> a variety of <u>subject areas</u>. It's often a <u>key</u> part
   of the <u>design process</u>, and is involved in <u>all</u> of the <u>strategies</u> mentioned above.

Iterative learning — look, cover, write, check... and again...
You need to know about the different approaches to design. It's the iterative one that you'll use during your NEA.

Design Strategies

2) You'll need to <u>work with different groups</u> of people to make sure your design is the best it can be.

Client Involvement

1) If you're working with a client, you may find it helpful to show them your <u>designs</u>, <u>models</u> and <u>prototypes</u> early on in the design process, to make sure it fits the <u>image</u> they had in mind.

2) If it's <u>not</u> what the <u>client</u> imagined, it offers them the opportunity to give <u>feedback</u> to <u>improve future iterations</u> of the design.

3) Each client will be <u>different</u>. Some may want to take a <u>hands-on approach</u> and be heavily involved in the design process, whereas others may <u>collaborate less</u> and leave it more to the designers.

User Involvement

1) This involves asking a <u>sample</u> of the <u>target market</u> for input in the design process.

2) The aim is to get feedback from <u>potential users</u> on your <u>design ideas</u>, <u>models</u> and <u>prototypes</u>, and make improvements so that the end product is more <u>appealing</u> to the <u>target market</u>.

Expert Opinions

1) Experts are other <u>professionals in the industry</u>, e.g. seamstresses, electronics experts etc.

2) You can <u>benefit</u> from their <u>experience</u> by asking them to look at your design and <u>suggest improvements</u>. For example, if you're making an <u>evening dress</u>, a seamstress may be able to suggest ways of altering a pattern so it's <u>cheaper</u> and <u>easier to make</u>, but still <u>looks good</u>.

Don't Get Stuck on a Bad Idea...

1) When you're designing a product it's easy to get <u>stuck on a particular idea</u> — for example, it may be an idea that is <u>similar</u> to an <u>existing product</u>, or a design that you've thought of <u>before</u>. This is called <u>design fixation</u> — it can stop you from thinking <u>creatively</u> and coming up with an <u>innovative design</u> idea.

2) <u>Following</u> a <u>design strategy</u> can help to <u>avoid design fixation</u> — they encourage you to look at your design in a <u>critical way</u> and make <u>improvements</u> where necessary.

3) There are other ways to help <u>avoid</u> design fixation. These include:

- <u>Collaboration</u> (see above and the previous page) — <u>brainstorming ideas</u> with <u>other people</u> and getting <u>honest feedback</u> on your design ideas might allow the <u>fixation</u> to be <u>broken</u>. It may help you to think more <u>creatively</u> too.

- <u>Focusing on new solutions</u> to the design brief. If you focus too much on existing products, it can be easy to attempt to <u>copy</u> them. It's great to use them as <u>inspiration</u>, but your aim should be to create a <u>new</u> and <u>unique</u> product, so make sure your ideas take a <u>fresh approach</u> to your product's <u>specific requirements</u>.

Brainstorm — no lightning but hopefully some striking ideas...

Q1 Which design strategy focuses on breaking down the design process into several different stages?

    a) Iterative design   b) User-centred design   c) Systems approach   [1 mark]

Q2 a) One of the first stages in the iterative design process is to make a prototype.
      Describe the stages that come after this.   [3 marks]

    b) Iterative design is a circular process. Explain what is meant by this.   [1 mark]

Q3 Give one reason why it's important to collaborate with the client early on in the design process.   [1 mark]

Q4 a) In design, what is the name for getting stuck on a particular idea?   [1 mark]

    b) How can following a design strategy help you to avoid getting stuck on one idea?   [1 mark]

Exploring and Developing a Design Idea

The first part of the <u>iterative design process</u> (see p.104) is <u>developing</u> a <u>design</u>. This is all about going from some <u>roughly sketched ideas</u> to your <u>first design</u> that you want to be made into a <u>prototype</u> (see p.114).

Detailed Sketches Help You Work on Finer Points of the Design

1) Your <u>initial sketches</u> will probably have been <u>rough</u>, <u>freehand</u> pencil drawings.

2) Trying out some <u>more detailed sketches</u> is the next stage.

3) It helps you to see what will actually <u>work</u> in practice and it might help you decide on <u>details</u> you hadn't thought about before, e.g. the sizes or positions of components, or how parts should be <u>constructed</u> and <u>fitted together</u>.

See pages 108-111 for lots of different drawing techniques.

Use Modelling to Improve Your Design

Models are often scaled-down (less than full-size) versions of a design.

1) Modelling is just making <u>practice versions</u> of your design, or parts of your design. It's a good way to <u>visualise</u> your design in <u>3D</u>, and to spot (and solve) any <u>problems</u>.

2) The modelling stage should also be used to try out <u>different materials</u> and <u>joining techniques</u>, and think about whether you could <u>reduce</u> the number of <u>parts</u> to make construction easier.

3) You can make models using materials that are <u>easy</u> and <u>quick</u> to <u>work with</u>, e.g. <u>cardboard</u>, <u>balsa</u>, <u>jelutong</u> (an easily workable wood), <u>newspaper</u> (used in textiles) or <u>foam core board</u>.

4) You can also use <u>construction kits</u> — these have different sized and shaped parts for you to build with.

5) Some types of modelling are only used for a <u>specific type</u> of <u>product</u>:

Toile

1) <u>Toiles</u> are <u>early versions</u> of a <u>clothing</u> design. They can be <u>worn</u> by a <u>model</u> or put on a <u>mannequin</u>, and they're used to work out the <u>proportions</u> and <u>fit</u> of the garment. Toiles for <u>one-off garments</u> are often fitted to the <u>intended wearer</u>.

2) The <u>fabric</u> used to make the toile should:
   - Be <u>cheap</u> — so the design can be <u>experimented</u> with and <u>improved</u>, without costing too much and <u>wasting</u> the <u>real fabric</u> you're planning to use.
   - Be <u>lightly coloured</u> — so you can <u>mark up</u> any <u>changes</u> needed onto it.
   - Have <u>similar properties</u> (e.g. weight, stretchiness, etc.) to the <u>real fabric</u> you're planning to use — this makes sure the <u>drape</u> of the garment can be <u>perfected</u> too, as it will be <u>similar</u> to when the real fabric is used.

3) <u>Calico</u> (an <u>unbleached cotton</u>) is a commonly used toile fabric.

Breadboards

1) Breadboards are used to <u>test</u> whether a <u>circuit design works</u>.

2) They are <u>boards</u> with <u>rows of holes</u> that <u>electronic components</u> can be <u>pushed through</u>. Wires are plugged into other holes to <u>complete the circuit</u> for testing.

3) Breadboards are useful as they <u>don't require soldering</u> (see p.91) — this is important when <u>developing</u> a circuit, as components can <u>easily be removed</u> and the <u>circuit easily changed</u>.

6) <u>CAD/CAM</u> (see p.4-5) can be used for <u>modelling</u> too. CAD can be used to draw a <u>detailed 3D design</u> and then <u>CAM machines</u> such as <u>3D printers</u> can use this design to produce a <u>model</u>.

7) <u>Mathematical modelling</u> is another way to model. It can show how an <u>object</u> or <u>system</u> will behave in <u>reality</u>. These models are made using <u>data</u> and information about <u>known relationships</u> between variables (factors that change).

Mathematical models can display information in a numerical or graphical form.

Multi-purpose breadboards — designing circuits over breakfast...

There are loads of different ways to make models of your design for testing and evaluating, but they all have a few things in common. The materials must be easy and quick to work with, and the model must be functionally and/or aesthetically similar to the final product so that the results from testing are realistic and useful.

Exploring and Developing a Design Idea

Test and Evaluate Each Model

1) After you've made the first model, do some <u>tests</u> to check that it's how it should be.

2) You'll probably find there are some things that <u>don't work out</u> quite how you'd hoped:

The model cupboard <u>looked great</u>...

...but the <u>doors</u> just <u>couldn't be opened</u>.

3) <u>Write down</u> what the problem is, and suggest how to <u>fix it</u>, e.g. use a different type of hinge. It might be a good idea to <u>mark up</u> the <u>changes</u> required <u>on the model itself</u> (e.g. as you would with a toile).

4) Record how the design develops — <u>take photos</u> of your models.

5) You should also <u>evaluate</u> each model against the <u>design specification</u> (see p.99). Take each point on the specification and see if your model is up to scratch.

6) You might end up having to <u>modify</u> your <u>specification</u> after you've evaluated your models.

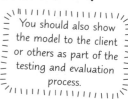

You should also show the model to the client or others as part of the testing and evaluation process.

Keep Going Until Your Model Meets All of the Design Criteria

1) You might find that you end up <u>changing something</u>, then <u>trying it out</u>, then making <u>another</u> change, and so on. That's just the way it goes sometimes.

2) Here's a summary of the process:

Make a model → Test and evaluate the model → Come up with ideas to improve the model

A very similar process is used to develop a prototype (see p.114) — the difference is that prototypes are testing the <u>real materials</u> and the actual <u>manufacturing processes</u> that are intended to be used to make the final product.

3) Throughout this process, you should be referring to the <u>design specification</u>.

Now You Should Be Ready to Start Making Prototypes

Once you've finished developing your ideas and have a <u>final design</u>, you're <u>ready</u> to start <u>developing</u> a <u>prototype</u> (see p.114). You'll need to have an idea of:

Ideas only need to be rough at this stage, as they'll be tweaked quite a lot during prototype development.

1) The best <u>materials</u>, <u>tools</u> and other <u>equipment</u> to use (and their availability). This might include <u>standard components</u> (see p.50).

2) The <u>assembly process</u> — this is important when it comes to <u>planning production</u> (see p.113).

3) The approximate <u>manufacturing time</u> needed to make each product.

4) How much it should <u>cost</u> to manufacture each product.

I feel like I'm going round in circles — it's making me dizzy...

Q1 a) What is meant by modelling? [1 mark]

b) Why is modelling useful when developing a design? [2 marks]

Q2 What is a toile and what is it used for? [2 marks]

Q3 Why are breadboards useful in developing a circuit design? [2 marks]

Q4 Describe what is meant by evaluating a model. [1 mark]

Q5 Once a final design has been developed, what is the next stage in the design process? [1 mark]

Drawing Techniques

To get your design across, you're going to need to draw it on paper. Here are a few techniques to help you communicate your design in the best way possible. Remember, practice makes perfect...

You Can Develop Design Ideas with Freehand Sketches

1) 'Freehand' means drawing without using any equipment (except a pencil or pen).

2) It's the quickest method of drawing, so it's handy for getting initial design ideas down on paper.

3) You can combine 2D and 3D sketches to explain details.

4) And you can annotate your sketches (add notes) to explain details further, e.g. describing the materials and processes you'd use.

Sketches aren't supposed to be perfect — they only need to get your ideas across.

Ideas for a space themed birthday card

Perspective Drawing Uses Vanishing Points

1) Perspective drawing tries to show what something actually looks like in 3D — smaller in the distance, larger close up. It does this by using lines that appear to meet at points called vanishing points.

2) These points are in the distance on the horizon line.

One-Point Perspective — for drawing objects head on.

1) Mark one vanishing point.
2) Draw the front view of the object head on.
3) Then draw lines to the vanishing point.

Two-Point Perspective — for drawing objects edge on.

1) Draw a horizon line horizontally across the page.
2) Mark two vanishing points near the ends of the horizon line.
3) Draw the object by starting with the front, vertical edge and then projecting lines to the vanishing points.
4) Remember that vertical lines remain vertical and all horizontal lines go to the vanishing points.

Vanishing point — Horizon line

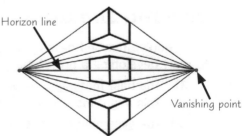
Horizon line — Vanishing point

Isometric Drawing Shows Objects at 30°

1) Isometric drawing can be used to show a 3D picture of an object.

2) It doesn't show perspective (things don't get smaller in the distance), but it's easier to get across the dimensions than in perspective drawing.

3) There are three main rules when drawing in isometric:

- Vertical edges are drawn as vertical lines.
- Horizontal edges are drawn at 30° (from horizontal).
- Parallel edges appear as parallel lines.

This drawing's been done on isometric dot paper. You could use plain paper and a 30°/60° set square instead.

Section Eight — Designing and Making

Drawing Techniques

Some <u>diagrams</u> are drawn <u>specifically</u> for <u>electronic</u> and <u>mechanical systems</u>...

System Diagrams are Flowcharts

1) System diagrams are <u>flowcharts</u> that <u>separate</u> a system into <u>input</u>, <u>process</u> and <u>output boxes</u>. This is useful when <u>developing</u> the <u>basic design</u> for a <u>system</u>.

2) <u>Annotations</u> are used to outline the <u>concept</u> of the <u>system</u>, <u>without</u> going into <u>detail</u> on how exactly you'd go about <u>making it work</u> — this is left to the <u>schematic diagrams</u> (see below).

3) Annotations should include the <u>components</u> or <u>mechanisms</u> that will be used at <u>each stage</u>, and they should <u>describe</u> what the components and mechanisms will <u>do</u>. They should be kept <u>brief</u> and <u>to the point</u> — they don't need to be very detailed.

Another example of a system diagram is the electronic egg timer on p.24.

For example — a <u>jack-in-the-box</u>:

INPUT	PROCESS	OUTPUT
A <u>crank</u> on a music box, inside a wooden box, is <u>turned</u>.	As it is turned, the <u>latch</u> to the box is <u>pushed open</u> by an arm.	A <u>jack</u> on a spring <u>jumps out</u> of the box.

4) <u>Freehand drawings</u> can be <u>added</u> to system diagrams to help get the point across, but it's really the <u>annotations</u> that are the <u>most important</u>.

Schematic Diagrams Show the Layout of a System

1) <u>Schematic diagrams</u> clearly show the <u>layout</u> of <u>electrical</u> and <u>mechanical</u> systems.

2) A <u>circuit diagram</u> is an example of a basic schematic diagram — it clearly shows how the <u>components</u> are <u>connected up</u>.

3) <u>Repair manuals</u> often use schematics to show the user how the <u>system</u> should be <u>assembled</u>.

Schematics are generally drawn later in the design process than system diagrams. For example, you need a basic flowchart outlining an electronic system before you can draw an appropriate circuit diagram for it.

4) Schematics are designed to be <u>easy to read</u> and <u>simple to draw</u>. For example:

- They <u>aren't drawn to scale</u> (see next page) — so components can be nicely <u>spaced out</u>. This means they <u>won't</u> contain <u>information</u> on any <u>dimensions</u> though.
- They often use <u>symbols</u>, e.g. <u>components</u> in a circuit diagram.
- They also can use <u>simplified</u> and often <u>unrealistic graphics</u>, e.g. a whole car schematic may show the <u>engine</u> simply as a <u>box</u>.

Circuit diagram

From my perspective, I wish exams stayed on the horizon...

In the exam, you may be asked to design something. Even if you're not great at drawing, you can still get a lot of the marks by annotating your drawings with comments that help to explain the details of your design.

Q1 What is freehand drawing? [1 mark]

Q2 What do perspective drawings try to show? [1 mark]

Q3 Give one advantage and one disadvantage of using isometric rather than perspective drawing. [2 marks]

Q4 Give one feature of schematic diagrams that makes them easy to read and understand. [1 mark]

More on Drawing Techniques

Here are a few more <u>drawing techniques</u> to help you along. First off, it's drawing things to <u>scale</u>...

Scale Drawings are Used to Draw Big Things (but smaller)

1) To draw a <u>big object</u> on a small piece of paper, you have to <u>scale it down</u>.

2) The object's still drawn in <u>proportion</u> — it's just <u>smaller</u>.

3) The <u>scale</u> is shown as a <u>ratio</u>. For example:

- A scale of <u>1:2</u> means that the <u>drawing</u> is <u>half the size</u> of the <u>real object</u>.
- A scale of <u>1:4</u> means that the drawing is <u>a quarter of the size</u> of the real object.
- Anything drawn at <u>1:1</u> is <u>full sized</u>.

Scale drawings are said to be "drawn to scale".

4) You can also <u>scale things up</u>. A scale of <u>2:1</u> means the <u>drawing</u> is <u>twice the size</u> of the real object.

5) The <u>scale</u> needs to be <u>clearly</u> shown on the diagram. It's a ratio, so it <u>doesn't have any units</u>.

This scale drawing...   ...is half the size of this toy submarine.

scale 1:2

Lines on a scale drawing should be labelled with the <u>lengths</u> of the <u>real object</u> — not the lengths of the lines on the paper.

EXAMPLE:

A jet ski is 1.2 m tall. A scale drawing of the jet ski has a height of 40 cm. What is the scale of the drawing?

First, convert the measurements to the same units.

Scale drawing height = 40 cm   Jet ski height = 1.2 m = 120 cm

Write the measurements as a ratio — "scale drawing : real object" = 40 : 120

Simplify the ratio by dividing each side by the same number.

Both sides will divide by 40.

40 ÷ 40 = 1 and 120 ÷ 40 = 3, so the scale of the drawing is <u>1 : 3</u>

6) To <u>check</u> you've scaled an object down properly, <u>measure</u> the lengths of the lines in your <u>drawing</u>. If you <u>multiply</u> those lengths <u>by the scale</u>, you should get the dimensions of the <u>real object</u>.

Exploded Diagrams Show How Parts Fit Together

<u>Assembly drawings</u> show how <u>separate parts join together</u>. <u>An exploded diagram</u> is a type of <u>assembly drawing</u>.

This exploded view is also an isometric drawing.

1) <u>Exploded diagrams</u> are always in <u>3D</u>.

2) You draw the product with <u>each separate part</u> of it <u>moved out</u> as if it's been exploded.

3) Each part of the product is <u>drawn in line</u> with the part it's attached to.

4) <u>Dotted lines</u> show where the part has been <u>exploded from</u>, and therefore where it <u>fits</u> into the overall product.

Exploded diagrams are designed so you can use them <u>on their own</u> — you <u>don't need many words</u> to explain how to <u>assemble</u> something. This means that as well as being sent to <u>manufacturers</u>, they're often used for <u>flat-pack furniture</u> instructions.

More on Drawing Techniques

Once you've _perfected_ your _design idea_, you'll need to produce an _accurate working drawing_ so that the _manufacturer_ can _make it_. Here's how to wow them (and the examiners) with some _super clear_ drawings...

Orthographic Projection Shows 2D Views of a 3D Object

1) _3rd angle projections_ are used very widely in industry to help the _manufacturer_ understand the design.

2) They show a _3D object_ as a set of _2D drawings_ viewed from _different angles_ — a _front_ view, a _plan_ view (as seen from _above_) and an _end_ view (as seen from the _side_).

3) The _symbol_ for _3rd angle orthographic_ projection is: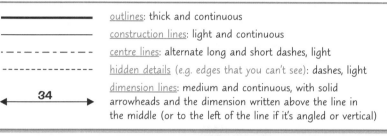

4) Each 2D view is drawn _accurately to scale_.

5) The _dimensions_ are always given in _millimetres_.

6) To avoid confusion, lines and dimensions must follow certain _conventions_:

————————	_outlines_: thick and continuous
————————	_construction lines_: light and continuous
—·—·—·—·—	_centre lines_: alternate long and short dashes, light
— — — — —	_hidden details_ (e.g. edges that you can't see): dashes, light
◄——34——►	_dimension lines_: medium and continuous, with solid arrowheads and the dimension written above the line in the middle (or to the left of the line if it's angled or vertical)

7) There's always a _gap_ between the _projection lines_ and the _object_.

8) The diameter of a _circle_ is shown by the symbol ∅ and an arrow inside the circle.

9) To _draw_ a _3rd angle orthographic projection_, you:

> 1) Draw the _front view_.
> 2) Add _construction_ and _centre_ lines to the _right_ — these help to draw the _outlines_ of the _end view_, which you should now complete.
> 3) Add _construction_ and _centre_ lines going _up_ from the _front view_.
> 4) Draw a line at _45°_ (as shown in the diagram above) in the _top right hand corner_ of the _front view_.
> 5) Draw _construction_ and _centre lines_ going _up_ from the _end view_. When they _reach_ the _45° line_, draw them going _90°_ and to the _left_ — these lines and the ones coming up from the front view can be used to draw the _plan view_.
> 6) Now the projections have taken shape, it should be easier to figure out which edges are the _hidden details_ in each of the views — _mark these_ on the drawings (see below).
> 7) Add the _dimensions_ — don't forget to use _millimetres_.

You may be asked to draw the third view when given the others. Draw construction and centre lines from the two views you have (using the 45° line if you need to project between plan and end views), and use these lines to draw the third view.

10) Some objects have _hidden details_ that you can only see from _certain views_. These can be shown in orthographic projections. For example:

A _box_ with a _hole_ through it...

You can see the hole from this view...

...but you can't see it in these views, so it's shown as a hidden detail.

Working drawings — even graphics have to earn a living...

Q1 The dimensions of a wooden box are 60 × 100 × 20 mm. Jo is doing a scale drawing of the box using a scale of 1:4. What should the dimensions of the drawing be? [1 mark]

Q2 In an exploded diagram, how are separate parts drawn to show how they join together? [2 marks]

Manufacturing Specification

A picture might be worth a thousand words, but working drawings don't give the manufacturer everything they need. The manufacturing specification provides the extra information required to make the product...

There are Loads of Things that Manufacturers Need to Know

A manufacturing specification can be a series of written statements, or working drawings and sequence diagrams (see next page). It has to explain exactly how to make the product, and should include:

1) clear construction details explaining exactly how to make each part,
2) materials — which materials to use for each part and how much will be needed,
3) equipment — what's needed at each stage,
4) sizes — precise dimensions of each part in millimetres,
5) tolerances (see p.40) — the maximum and minimum sizes each part should be,
6) finishing details — any special information, such as 'laminate the paper with aluminium',
7) quality control instructions (see p.40) — what needs to be checked, and how and when to check it,
8) costings — how much each part costs, and details of any other costs involved.

Working drawings give measurements, materials and construction details (e.g. where to make folds).

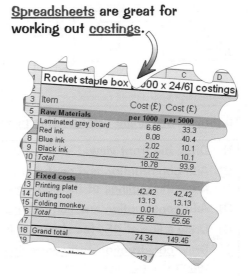

Spreadsheets are great for working out costings.

Each Stage Needs to be Planned in Detail

Take each stage of the manufacturing process and plan it in detail. You need to think about:

1) how long each stage will take,
2) what needs to be prepared before you can start each stage,
3) how you will ensure consistency and quality, e.g. using jigs, formers and measuring tools,
4) how you will do quality control checks,
5) what health and safety precautions you will have to take to be safe when making your product.

Remember, your methods would probably change if you were going to produce your design in quantity.

Sorry — puns weren't specified...

As pretty as your working drawings may be, you can't just give a manufacturer a picture and expect them to bring your design to life. Make sure you know everything that should be included in the manufacturing specification.

Section Eight — Designing and Making

Manufacturing Specification

Making a <u>few</u> examples of your product is (relatively) <u>easy</u>. But <u>mass-producing</u> it is much harder and takes a lot of careful <u>planning</u>. This is where <u>sequence diagrams</u> such as <u>flowcharts</u> and <u>Gantt charts</u> come in handy...

Charts Help to Plan the Manufacturing Process

Work Order

1) A <u>table</u> or <u>flowchart</u> can be used to plan the <u>work order</u> — the <u>sequence</u> in which tasks should be carried out.

2) They can also include <u>tools</u>, <u>quality control</u> stages, <u>safety</u>, and so on.

Start and end a flow chart with a sausage-shaped box.

<u>Processes</u> (the things you <u>do</u>) go in rectangular boxes.

<u>Decisions</u> go in diamond-shaped boxes. These let you show where <u>quality</u> should be <u>checked</u>.

Prototype staple box			
Day	Process	Tools needed	Quality ch
1	Print designs	Airbrush, pens, dry transfer lettering	Make s
2	Cut out net	Scalpel, metal rule	Chec
	Score and fold net	Scissors, metal rule	

The diamond-shaped boxes show where you will stop and see if your product looks and works how it should. If you find it doesn't, go back and make sure it's done properly before you move on.

Gantt Charts

1) You also need to work out <u>how long</u> each stage will take, and how these times will fit into the <u>total time</u> you've allowed for production. One way to do this is with a <u>Gantt chart</u>.

2) The <u>tasks</u> are listed down the <u>left-hand</u> side, and the <u>timing</u> is plotted across the <u>top</u>.

3) The <u>coloured squares</u> show <u>how long</u> each task takes and the <u>order</u> they're done in. In this chart, each square represents 5 minutes.

You can start this next stage while the paint is drying, so you can draw the bars underneath each other — they can happen at the same time.

Make sure you know how to read the timings on a Gantt chart — the times at the top show how much time has passed in the overall production process — not how long each stage will take. For example, 55 minutes into production, the writing should be being applied, and this should take about 10 (not 55-60) minutes to complete.

Just charting my revision — eat, sleep, revise, repeat...

Q1 a) What is a manufacturing specification?   [1 mark]
      b) Give three things that a manufacturing specification should include.   [3 marks]

Q2 On a work order flowchart, how would you show where quality control should take place?   [1 mark]

Q3 a) In the Gantt chart above, why don't many of the bars overlap?   [1 mark]
      b) How long does the whole production process last?   [1 mark]
      c) Give the longest stage, and state how long that stage should take to complete. [2 marks]

Developing Prototypes

Next up, it's developing prototypes. I know it sounds a bit dull, but this is super useful for your non-exam assessment (NEA, see p.1), where you'll be developing a prototype of your own, so it's well worth a read...

Prototypes are Full-Size, Working Products or Systems

1) Prototypes are a step further than the models on page 106 — they're full-size, fully-functioning products or systems.

2) They're made using the materials and manufacturing methods that are intended for the final product.

3) This allows you to test the product and the production methods you're going to use to make sure they're as you want them.

The final product you make in your project is a prototype of your design.

Prototypes Allow You to Evaluate the Manufacturing Process

1) Making a prototype allows you to check that the manufacturing specification (see p.112) is correct.

2) Any problems identified during this process will need to be solved, and the manufacturing specification modified to include the changes.

3) Making a prototype can also help you calculate some of the manufacturing costs — this is important so you don't end up with a final product that costs far too much to make. In industry, these costs can include:

• Materials and components	• Cost of new equipment
• Labour	• Energy
• Packaging	• Waste disposal

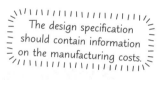
The design specification should contain information on the manufacturing costs.

You Can Evaluate Your Design Using Your Prototype

1) Prototypes are evaluated by checking that they meet the criteria set out in the design specification (see p.99). It's best to go through these criteria one by one.

2) You'll probably find that parts of your design don't work out the way you wanted them, and so don't fit the design specification. If so, you need to think about the improvements you could make. You need to be able to reject the part of the design that doesn't work and justify an alternative. For example:

Remember, the design specification contains the conditions a product should meet, e.g. the product should be functional, safe to use, have good aesthetics, etc.

Design Specification for a Reusable Shopping Bag
- The finished bag must be retailed for £3 or less.
- The bag must be made out of environmentally-friendly material.
- It must be able to hold 4 kg of stuff.

Maybe the plain seams (see p.80) attaching the handles to the bag aren't strong enough to hold the 4 kg weight. Instead, you could use a flat-felled seam to attach the handles to the bag, as this is a stronger and more durable method of attachment.

In your project, you'll be asked to evaluate your prototype, so this stuff should come in handy there too...

3) You also need to make sure that the prototype meets the requirements of the design brief — what the client wants or needs (see p.98).

Developing Prototypes

You Should Also Get Feedback from Other People

1) Prototypes are useful as they allow other people to try out your product.

2) Feedback from the client, potential customers and experts in the industry (see p.105) provide additional suggestions on how you could modify your design.

3) This will hopefully improve the product so that it's marketable — appealing to the target market, is fit for purpose and will sell (hopefully).

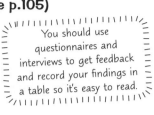
You should use questionnaires and interviews to get feedback and record your findings in a table so it's easy to read.

> E.g. potential customers might test the reusable shopping bag prototype and say that the cord handles are good for carrying a load of shopping, but when the bag is empty, the cords make it bulky and difficult to fold away.
>
> This means you can justify making the handles out of flexible cotton straps instead, although a prototype with these should be tested and evaluated to make sure there aren't other problems with these.

Cotton fabric straps instead of cord.

Keep Going Until You Get it Just Right

1) You might find that you end up modifying something, then evaluating it, then making another modification and testing that, and so on. This repetitive process is known as iterative design (see p.104) and is all part of developing a good product.

2) Depending on your time and resources, you could make any number of alternative prototypes. Once you've evaluated them all, go back and select the best one.

3) When you're doing this for real with your own product, you need to keep a record of what you find out from each prototype and the changes you make to your design as a result. Include your justification too, e.g. why each new prototype better fits the specifications.

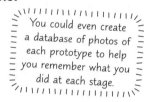
You could even create a database of photos of each prototype to help you remember what you did at each stage.

4) Once a product is as perfect as it can be, it's time to consider production on a larger scale.

5) Testing prototypes is very important in industry, as any mistakes that aren't picked up will cost a lot of money to put right.

Goldilocks knew how to prototype — beds, porridge 'n' all...

Developing a really good prototype involves looking at each prototype you make critically and making improvements. Sometimes you can take one step forward and two steps back if your changes mean a different point on the design specification isn't met, so think about any knock-on effects a change might have.

Q1 What are prototypes? [2 marks]

Q2 Give two reasons why it's useful to make prototypes using the manufacturing specification. [2 marks]

Q3 Sam is making a prototype for a winter coat designed for the mass market.

    a) Suggest two groups of people he could ask for feedback on his prototype. [2 marks]

    b) Explain why getting feedback on his prototype could be useful. [2 marks]

Q4 Tammy makes a prototype pyjama case in the shape of a dinosaur, with a button fastening for the mouth. When she asks her friends what they think about it, they say that the buttons don't look much like teeth. What could she do next?

[1 mark]

Using Materials Efficiently

Efficient use of a piece of material means making as many parts or products as possible with it...

Wasting Materials Means Losing Money

1) Any material that manufacturers waste means a loss of money — so when they're making products, they have to make the best use of the materials.

2) If manufacturers can make their product for less money, it means they can make more profit or afford to sell it to us more cheaply.

3) Careful planning of where to cut (nesting — see next page) and accurate marking out can reduce waste.

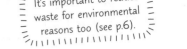
It's important to reduce waste for environmental reasons too (see p.6).

There are Loads of Different Tools for Marking Out

1) Marking out is making a mark so you know where to cut, drill or assemble your material.

2) This helps to make sure that products or parts are made accurately, consistently and to a high quality.

3) Marking out may be as basic as using a pencil on a piece of wood (you shouldn't use a pen as it seeps into the wood and you can't get it out) or a felt tip on some plastic — but there are lots of other ways:

SCRIBER — used like a pencil but it scratches a mark into metal and plastic.

ENGINEER'S BLUE and **MARKING BLUE** are dyes you put on metal so any scribe marks show up better.

PATTERN MASTER — used to help you draw paper patterns. The curved edges are for marking out smooth curves and the parallel lines are for marking out extra width for seam allowance.

SLIDING BEVEL — can be set at any angle to guide a marking knife.

ODD-LEG CALIPER — marks a line parallel to an edge.

TAILOR'S CHALK — used to transfer markings onto a fabric that you can remove later.

MARKING KNIFE — used to score lines in wood (it cuts the fibres to stop them splitting during sawing).

MARKING GAUGE — scratches a line in wood parallel to an edge.

Things need to be cut or shaped to the correct size — so measure carefully when you're marking out. Steel rules can be used to measure lengths. Combination squares can be used to check things are level, measure angles, and find the centre of circles.

TRY SQUARE — helps to accurately mark out right angles (90°).

TEMPLATES and **PATTERNS** (see p.42-43) — templates can be drawn round to mark out the same shape. Patterns can be cut round.

4) You can also mark out with the help of reference points, lines and surfaces, and coordinates (see p.42). You don't always need to mark out though — jigs (see p.43) can be used to guide tools instead.

When Marking Out, Remember...

Waste side
Marking line
Wood you want to keep — cut slightly too big

1) Always measure things out twice — that way, you're less likely to make a mistake.

2) Keep the marks as thin as possible — otherwise your cutting might be inaccurate.

3) Remember to get rid of the marks when you've finished so it doesn't look untidy.

4) If possible, mark out on a surface that the user won't see — for example on the back of a door, or the inside surface of a box.

5) When you cut, always do it on the waste side of the line so it doesn't end up too small. If it's a bit big, you can always file it down to size — but if it's too small you'll have to cut it again.

6) And remember, you can't mark and cut everything perfectly — nobody's perfect. But it's important to work within the required tolerance (see p.40).

Section Eight — Designing and Making

Using Materials Efficiently

Arranging Things Efficiently Helps to Reduce Waste

1) When <u>batch</u> or <u>mass producing</u> a <u>product</u> (see p.38 and 39), the <u>same</u> shapes are <u>cut</u> out <u>again and again</u> often from stock forms.

2) Waste can be reduced by making as <u>many products</u> as possible from the <u>least amount of material</u>. For example, manufacturers can try to cut as many shapes from <u>one sheet</u> of material as possible.

3) Some shapes will <u>tessellate</u> — this means that <u>repeats</u> of the shape <u>fit together</u> without any <u>gaps</u> or <u>overlapping</u> pieces, e.g. a hexagon. These shapes can be cut from sheets of material with <u>very little waste</u>.

4) When <u>planning</u> the cutting and shaping of a material, manufacturers try to come up with the most <u>efficient arrangement</u> of shapes to <u>minimise waste</u> — this process is called <u>nesting</u>.

5) To <u>work out</u> which <u>arrangement</u> is <u>best</u>, you can work out the <u>area</u> (if it's a <u>sheet</u>) or <u>volume</u> of <u>material</u> that is <u>going to waste</u>:

- First, calculate the <u>area</u> (or <u>volume</u>) of <u>one shape</u>.
- <u>Multiply</u> this by the <u>number of shapes</u> you can <u>fit</u> onto the <u>material</u> — this gives the amount of material that <u>can be used</u>.
- Work out the <u>area</u> (or <u>volume</u>) of <u>material</u> you <u>started with</u>.
- Then <u>subtract</u> the amount of <u>useful material</u> from this — this leaves the amount of <u>waste material</u>.

The arrangement with the <u>smallest amount of waste</u> should be <u>chosen</u>.

Don't worry — you're allowed to use a calculator in the exam...

You can fit a fair few cube nets onto this sheet of material if you arrange them right.

 EXAMPLE: The triangle on the right can fit onto a 21 cm × 29.7 cm sheet of paper, 23 times. Calculate the amount of paper wasted in a single sheet.

Area of a triangle = ½ × width × height = ½ × 6 × 7.5 = 22.5 cm²
Total area of triangles = 22.5 × 23 = 517.5 cm²
Area of paper = length × width = 21 × 29.7 = 623.7 cm²
So the amount of paper wasted = 623.7 - 517.5 = <u>106.2 cm²</u>

Remember to write your units, e.g. cm².

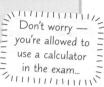

7.5 cm
6 cm

6) One way to <u>quickly work out</u> an <u>efficient layout</u> is to use <u>CAD</u> — e.g. CAD can be used in <u>lay planning</u> to find an efficient way of laying out <u>pattern pieces</u> onto a <u>sheet of fabric</u>.

7) Remember to include any <u>extra material</u> needed to <u>join</u> materials together when <u>marking out</u> your shapes.

- For example, <u>nets</u> used to make packaging (see p.53) have extra <u>flaps</u> of material that are <u>glued</u> to <u>another part</u> of the <u>packaging</u>. This <u>holds it</u> all <u>together</u>.
- In clothes, extra fabric is needed for the <u>seam allowance</u> (see p.80).
- In <u>joints</u> (e.g. in wood or metal), extra material is given so there can be an <u>overlap</u>.

How do you grade DT exams — with a marking gauge...

Marking out is pretty important. It's something you need to do accurately when hand-making prototypes. You'll end up wasting less time and materials, as you're more likely to get it right first time.

Q1   a) What is meant by marking out?   [1 mark]

       b) Suggest a piece of equipment that you could use to mark out on wood.   [1 mark]

Q2   Explain what it means if a shape tessellates.   [1 mark]

Q3   Explain what is meant by nesting.   [1 mark]

Q4   Bob wants to cut out 12 stars from a 70 × 50 cm sheet of board. Each star has an area of 150 cm².
How much waste board will he have when he's finished?   [3 marks]

Working Safely

You need to <u>think</u> about <u>safety</u> when using different <u>tools</u> and <u>equipment</u>.
This is <u>important</u> for your <u>own safety</u> but also the safety of the <u>people around you</u>.

Safety is Essential when you Make a Product

A lot of this is common sense, but it's <u>incredibly important</u> stuff so pay attention...

Wear Protective Clothing and Equipment

1) With <u>hazardous materials</u> wear a <u>face mask</u> or <u>goggles</u> and <u>protective gloves</u>. These will protect the <u>skin</u> and <u>eyes</u>, e.g. from <u>splashes</u> and <u>spillages</u>.
2) If the material is <u>hot</u>, wear <u>protective gloves</u> and an <u>apron</u>. For some jobs, e.g. welding, you should also wear a <u>tinted face shield</u>.
3) When using cutting machines, wear <u>chainmail gloves</u> to protect your hands.
4) When using <u>pins</u> and <u>needles</u>, use a <u>thimble</u>.
5) When using <u>noisy machinery</u>, wear <u>ear protection</u>.

Be Careful with Tools and Machinery

1) When working with tools and machinery, make sure you have your <u>sleeves rolled back</u>, <u>ties tucked in</u> or taken off, <u>apron strings tucked in</u>, <u>necklaces</u>, <u>watches</u> and <u>rings taken off</u>, and <u>long hair tied back</u>.
2) Never leave any machines <u>unattended</u> while switched on.
3) Never use a machine or hand tool unless you've been <u>taught</u> how to.
4) Know how to <u>switch off</u> machines in an <u>emergency</u>.
5) Don't <u>change parts</u> (e.g. drill bits) on a machine until you've <u>isolated</u> it from the mains.
6) Always <u>secure</u> work safely — e.g. clamp work securely before cutting.
7) Use <u>guards</u> on <u>machines</u> that have them, e.g. sanding machines and pillar drills.
8) Use a <u>dust extractor</u> if the process produces dust. (It might be sensible to wear a <u>dust mask</u> and <u>eye protection</u> too.)
9) <u>Carry</u> and <u>store</u> sharp tools safely.

Handle Materials and Waste Sensibly

1) <u>Choose</u> your materials sensibly — only use <u>hazardous materials</u> when absolutely <u>necessary</u>.
2) Make sure materials are <u>safe to handle</u>, e.g. <u>file down</u> rough edges before starting work.
3) When <u>storing</u> material, make sure it's <u>put away safely</u> so it can't fall and injure anyone. Keep <u>flammable liquids</u> away from <u>naked flames</u> and red-hot heating elements.
4) When using <u>toxic chemicals</u> such as dyes, finishes and solvents, you need adequate <u>ventilation</u> to avoid <u>inhalation</u> of vapours.
5) Dispose of <u>waste</u> properly so that it doesn't <u>harm</u> the <u>environment</u>.

CAUTION FLAMMABLE LIQUIDS

Most chemicals will have guidelines and regulations telling you why they're hazardous and how to use, store and dispose of them safely.

Careful — you could have someone's eye out with that...

Making a product is full of risks and hazards. Some can be avoided by being careful and sensible. However, you must pay attention to any labels or safety instructions on chemicals and equipment — they're there for a reason...

Working Safely

Risk Assessments are also Important

1) A <u>risk assessment</u> is used to <u>identify</u> and <u>minimise</u> any <u>risks</u> when <u>working</u>.

2) When writing a risk assessment, think about:

- What could be a <u>hazard</u>?
- What <u>precautions</u> could be taken to make sure the risk is minimised?

3) Here's part of a risk assessment for making a carved wooden box.

4) Risk assessments are <u>carried out</u>:

- On the <u>product</u> during the <u>design process</u>, to assess potential risks to the <u>end user</u>.
- By <u>manufacturers</u>, for the <u>materials</u>, <u>chemicals</u> and <u>equipment</u> that <u>workers</u> will have to use during the manufacture of the product.

Risk assessment for project

Hazard	Precaution — how to reduce the risk
Clothing could get caught in the sanding machine.	Tuck clothes in and wear an apron.
Fine dust created when using a sanding machine.	Wear a mask and use a dust extractor.
Fingers could be cut when using a craft knife.	Use a safety rule to protect fingers.

You Can Carry Out Tests to Ensure Safety

Even in school, there are things you can do to make sure the <u>product</u> you're designing is <u>safe</u>.

1) <u>Research your materials</u> carefully and <u>test them</u> to make sure they're suitable, e.g. see how far they'll bend, check whether they catch fire easily, etc.

2) Use <u>standard components</u> (see p.50) wherever you can, because these have already been tested by the manufacturer — this helps make sure that safety standards are met.

3) Make <u>prototypes</u> and carry out real-life <u>simulations</u>, e.g. if you're making a cot, check that a toddler's finger can't get trapped in the gaps between the bars (using a metal rod of about the same width as a toddler's finger, for example).

4) Get <u>electrical</u> items <u>PAT tested</u> (PAT stands for Portable Appliance Test). The test makes sure that portable electric products won't hurt anyone if used properly. School technicians can sometimes do this.

5) You can test products using <u>CAD</u> software. If you've selected the <u>right materials</u>, you can simulate <u>stress tests</u>, such as how well a product will stand up to an impact.

Daisy's definitely safe — she passes her cow PAT twice a day...

Q1 Suggest a piece of protective clothing or equipment that you would use when:

a) handling hot materials [1 mark]

b) using cutting machinery [1 mark]

Q2 Why do you need good ventilation when dealing with toxic chemicals? [1 mark]

Q3 a) What two things should be written down in a risk assessment? [2 marks]

b) Suggest why risk assessments are carried out on the product during the design process. [2 marks]

Q4 Explain how each of the following can be useful in making sure your product is safe for use.

a) standard components [1 mark]

b) prototypes [1 mark]

c) CAD software [1 mark]

Exam Technique

Here's What to Expect in the Exam

1) At the end of your GCSE Design and Technology course you'll have to sit an exam. It's worth <u>50%</u> of your <u>total mark</u>. The exam is worth <u>100 marks</u> and is <u>2 hours</u> long.

2) The questions in the exam will test you on <u>all areas</u> of the <u>course</u>. The exam is split into <u>three sections</u> which test you on different parts of the course. Each section contains different <u>types of questions</u>:

Section A
- Consists of a mixture of <u>multiple choice questions</u>, worth <u>one mark</u> each, and <u>short answer</u> questions.
- The multiple choice questions have one correct answer and three incorrect answers. Only pick <u>one</u> answer — if you pick more than one you won't get the mark.
- <u>20 marks</u> are available in <u>total</u>.

Section B
- Consists of a few <u>short answer</u> questions and one <u>extended response</u> question, which is worth more marks.
- Some questions may offer a <u>choice</u> on what to write about — e.g. a choice of <u>materials</u>, <u>products</u>, etc. It's worth spending a moment <u>reading</u> the <u>whole question</u> to figure out which option to choose — pick the one you could write the <u>best answer</u> for.
- <u>30 marks</u> are available in <u>total</u>.

Section C
- Consists of a mixture of questions including <u>short answers</u> and <u>extended response</u> questions.
- <u>50 marks</u> are available in <u>total</u>.

Remember These Tips

1) <u>Always</u> read every question <u>carefully</u>.
   Don't write an essay about modelling if it's asking about developing prototypes.

2) Use the number of <u>marks</u> as a <u>guide</u> on <u>how much</u> to <u>write</u>.
   Try to write a point for each mark. For example, if the question's worth 3 marks, write 3 points — not just a single sentence.

3) Write your answers <u>clearly</u>, using <u>good grammar</u>, <u>spelling</u> and <u>punctuation</u>.
   If the examiner can't read your answer, you won't get any marks, even if it's right.

4) Use the correct <u>terminology</u>.
   Know your technical terms — CMYK, MDF, PIC... you can't write about these if you don't know what they mean.

5) Make your <u>sketches clear</u>.
   Draw clearly and label your sketches — it makes it easier for the examiner to see what you're trying to get across.

6) <u>Don't panic</u>.
   If you really can't do a question, just leave it and move on to the next one. You can come back to it at the end.

Understand the Command Words

Questions will often use <u>command words</u> — these words tell you <u>how</u> to answer the question. If you <u>don't know</u> what they mean, you might not answer the question <u>properly</u>. Here are some you might come across:

State — You should give a <u>short answer</u> or <u>list</u> — you don't need to explain why.

Define — You should give a <u>clear</u>, <u>precise</u> meaning of the word or phrase.

Outline — You should give a <u>brief</u> <u>summary</u> of a process.

Explain — You should <u>give reasons</u> to show <u>why</u>.

Describe — You should give a <u>detailed</u> <u>description</u> of something.

Discuss — You should make a <u>balanced</u> <u>argument</u> covering a range of opinions.

Assess / Evaluate — You should use <u>evidence</u> and your <u>own knowledge</u> to come to a <u>conclusion</u>.

Exam Technique

Plan Out Your Answer to Extended Response Questions

Extended response questions are longer questions worth 6 or more marks and with a scary number of dotted lines underneath... They often use command words like discuss, evaluate or assess (see previous page).

1) Your answer must be well-written (good spelling, grammar and punctuation) and well-structured.

2) Before you start an extended response question, jot down the points you want to make and plan your answer to help structure it well and avoid repeating things.

3) You might have to weigh up the advantages and disadvantages of something, or cover both sides of an argument then form your own opinion.

4) Make absolutely sure you're answering the question and not just waffling on.

Here's an example:

'The only way that designers can design products that are environmentally friendly is by using recycled materials.' Discuss whether you agree with this statement. **[8 marks]**

Using recycled materials in a product is good for the environment, as it reduces the need for raw materials to be extracted from their source, which causes environmental damage. For example, the extraction of rocks and metals often requires a lot of land to be cleared for quarries and mines. Using recycled materials therefore reduces the need for this land clearance and preserves natural habitats that would otherwise have been destroyed. The extraction of raw materials can also consume a large amount of energy which, if provided by burning fossil fuels, causes the release of carbon dioxide (and other air pollutants) so contributes to global warming. Using recycled materials therefore reduces a product's carbon footprint. It also often means that the product can be recycled at the end of its life, which is more environmentally friendly than the product ending up in landfill.

However, there are many factors other than using recycled materials that a designer can consider to make a product that is environmentally friendly. Renewable materials such as wood (for timber, paper and board) that is obtained from a sustainable source can be chosen, rather than materials from sources that are unsustainable and damage the environment (e.g. those that result in deforestation). Designing a durable product, e.g. one that doesn't have built-in obsolescence, helps reduce the consumption of energy and materials used to make replacement products, and the pollution linked to this. Using biodegradable materials is also environmentally friendly as they rot away in landfill sites.

In summary, using recycled materials is beneficial for the environment, but it isn't the only way of designing an environmentally friendly product — there are several other factors that can be considered too.

A well-structured answer showing **detailed** knowledge is required to get all **8 marks**.

This explanation shows **good knowledge** of the impacts of fossil fuels on the environment.

You'll need to discuss **both sides** of the argument.

You'll need to use the **correct terminology**.

Include a **conclusion** that **answers** the **question** and **rounds off** your answer well.

This answer has **good grammar**, **spelling** and **punctuation**.

For **extended response** questions, **don't worry** if you don't **fill** all the **lines** — there are just plenty to make sure you have **enough space** to answer the question.

Glossary

alloy	A mixture of <u>two or more metals</u>, or a <u>metal</u> mixed with <u>one or more elements</u>.
anthropometrics	<u>Human body measurement</u> data.
automation	The use of <u>machines</u> to do a task <u>automatically</u> without much, or any, human input.
batch production	The <u>production method</u> used to make a <u>specific quantity</u> (a batch) of <u>identical</u> products.
CAD/CAM	Computer aided design/manufacture. <u>Designing</u> and <u>manufacturing</u> using a <u>computer</u>.
carbon footprint	The amount of <u>greenhouse gases</u> released into the atmosphere by <u>making</u>, <u>using</u> and eventually <u>reusing</u>, <u>recycling</u> or <u>disposing</u> of something at the end of its lifetime.
composite	A material made by <u>bonding</u> two or more different materials together.
continuous production	The <u>production method</u> used to make <u>large amounts</u> of a product <u>non-stop</u>.
corrosion	The <u>gradual destruction</u> of a material as it <u>reacts</u> with a substance, e.g. rusting of iron.
deforestation	Cutting down large areas of forest <u>without planting new trees</u> to replace the old ones.
design brief	The <u>instructions</u> that a <u>client</u> gives to a <u>designer</u> about what they want a product to <u>be like</u>.
design specification	A list of <u>criteria</u> that a product should meet.
ergonomic	A product that is <u>easy</u> and <u>comfortable</u> for people to <u>use</u>.
ferrous metal	A metal or alloy that <u>contains iron</u>.
fibre	A thin, <u>hair-like strand</u>. Fibres can be spun into yarn, or used as they are, to make fabrics.
filament	A <u>long continuous</u> length of <u>fibre</u>.
finite resource	A resource that will <u>run out eventually</u>, e.g. crude oil. Also called a <u>non-renewable</u> resource.
flexible manufacturing system (FMS)	A <u>set</u> of different <u>machines</u> which carry out the different <u>stages of production</u>. These <u>computer-controlled</u>, <u>automated systems</u> are designed to be <u>easy to adapt</u>.
hardwood	A type of wood that comes from <u>slow-growing trees</u> with <u>broad leaves</u> (mainly deciduous trees). It's usually <u>denser</u> and <u>harder</u> than softwoods.
integrated circuit (IC)	A <u>tiny</u>, <u>self-contained</u> circuit which can contain <u>billions of components</u>.
iterative design	A <u>design strategy</u> that involves constantly <u>evaluating</u> and <u>improving</u> a product's design.
knitted fabric	A fabric made from <u>yarns</u> held together by <u>interlocking loops</u>.
lean manufacturing	An approach to manufacturing that aims to <u>minimise</u> the <u>resources used</u> and <u>waste produced</u>.
manufactured board	A material made by <u>compressing</u> a mixture of <u>glue</u> and <u>processed pieces of wood</u> into <u>panels</u>.
manufacturing specification	A series of <u>written statements</u>, or <u>working drawings</u> and <u>sequence diagrams</u>, that tells the manufacturer exactly <u>how to make the product</u>.
market pull	When a product is made due to <u>consumer demand</u>.
market research	Asking the <u>target market</u> questions to find out their <u>likes/dislikes</u> (and so on) to help the designer understand what the target group <u>wants</u> from a product.
marking out	<u>Making a mark</u> in a material to show where it is to be <u>cut</u>, <u>drilled</u> etc.
mass production	The <u>production method</u> used to produce a <u>large number</u> of <u>identical</u> products on an <u>assembly line</u>.
microcontroller	A type of <u>integrated circuit</u> that can be <u>programmed</u>. Works as a <u>mini-computer</u>.
modelling	Making a <u>practice version</u> of a design or part of a design.

Glossary

modern material	A material that has been <u>developed</u> for a <u>specific application</u>. They are often developed through the <u>invention</u> of a <u>new process</u> or the <u>improvement</u> of an <u>existing one</u>.
natural fibre	A type of fibre that is harvested from <u>natural sources</u>, e.g. <u>plants</u> and <u>animals</u>.
non-ferrous metal	A metal or alloy that <u>doesn't contain iron</u>.
non-finite resource	A resource that can be <u>replaced</u> by <u>natural processes</u> as <u>fast</u> as it's <u>consumed</u> by humans, e.g. softwood trees in a plantation. Also called a <u>renewable</u> resource.
non-woven fabric	A fabric made from layers of <u>fibres</u> (not yarns) held together by <u>bonding</u> or <u>felting</u>.
one-off production	<u>Production method</u> used to produce a <u>single</u>, <u>unique</u> product at a time.
orthographic projection	A <u>2D scale drawing</u> of a 3D object showing the <u>front</u>, <u>plan</u> and <u>end views</u>.
planned obsolescence	When a product is <u>designed to become useless</u> quickly, e.g. a disposable razor.
printed circuit board (PCB)	A board with <u>copper tracks</u> that connect <u>components</u> in a circuit.
product analysis	<u>Examining</u> and <u>disassembling</u> a current product to get ideas for a new product or design.
production aid	A tool or technique used to <u>speed up</u>, <u>simplify</u> or help control the <u>accuracy</u> of a production process.
prototype	A <u>full-size</u>, <u>fully-functioning</u> product or system. It is built so that the <u>product</u> and <u>production methods</u> can be <u>evaluated</u> before the product is manufactured on a larger scale.
quality control	The <u>checks</u> that are carried out on a <u>sample</u> of materials, products or components throughout <u>production</u> to make sure <u>standards</u> are being <u>met</u>.
seasoning	The <u>drying</u> of wood to make it <u>stronger</u> and <u>less likely to rot</u> or <u>twist</u>.
smart material	A material that <u>changes its properties</u> in response to stimuli (a <u>change</u> in the <u>environment</u>).
softwood	A type of wood that comes from <u>fast-growing trees</u> with <u>needle-like leaves</u> (mainly evergreen trees). It's usually <u>less dense</u> and <u>softer</u> than hardwoods.
standard components	Common <u>fixings</u> and <u>parts</u> that manufacturers <u>buy</u> instead of manufacturing themselves.
staple fibre	A <u>short fibre</u>, e.g. cotton fibres are staple fibres.
stock forms	The different <u>shapes</u> that materials can be <u>bought in</u>.
sustainable	A sustainable process or material is one that can be used <u>without causing permanent damage</u> to the environment or <u>using up finite resources</u>.
synthetic fibre	A <u>man-made</u> fibre that is produced from <u>polymers</u>.
system	A collection of parts that work together to do a particular function. Made up of <u>inputs</u>, <u>processes</u> and <u>outputs</u>.
target market	The <u>group</u> of people you want to <u>sell</u> a product to.
technology push	When <u>advances in technology</u> drive the design of <u>new products</u> and the <u>redesign</u> of old ones.
thermoforming plastic	A type of plastic that can be <u>melted</u> and <u>remoulded</u> over and over again.
thermosetting plastic	A type of plastic that undergoes a <u>chemical change</u> when heated, which makes it <u>permanently hard</u> and <u>rigid</u>. Thermosetting plastics <u>can't be remoulded</u>.
tolerance	The <u>margin of error</u> allowed for a measurement of part of a product. Tolerances are usually given as an <u>upper</u> and <u>lower limit</u> e.g. 23 mm ($\pm$ 2).
woven fabric	A fabric made by <u>interlacing two sets</u> of <u>yarn</u>.
yarn	A thread made by <u>twisting</u> fibres together. Yarns are <u>woven</u> or <u>knitted</u> to make fabrics.

Answers

A note about answers and marks
The answers and mark schemes given here should be used mainly for guidance, as there may be many different correct answers to each question — don't panic if your answers are a bit different.

Section One — Key Ideas in Design and Technology

Page 3 — Technology in Manufacturing

Q1 A Just-in-Time (JIT) system is a system of stock control where the manufacturer gets the materials and components delivered as they're needed *[1 mark]* and uses them as soon as they're delivered *[1 mark]*.

Page 5 — Production Systems — CAD/CAM

Q1 computer aided design *[1 mark]*

Q2 a) They follow x,y,z coordinates to move their tools to cut out or build up a design *[1 mark]*.

b) E.g. a CNC milling machine / CNC router / 3D printer / laser cutter *[1 mark]*.

c) E.g. CAM machines that carry out subtraction remove material from a solid block to form an object *[1 mark]*. CAM machines that carry out addition add material to build up an object *[1 mark]*.

Q3 e.g. plastic/wax *[1 mark]*

Q4 E.g. designs produced by CAD can be sent electronically to be manufactured in a different part of the world near where they'll be sold *[1 mark]*. This means that companies can save on the cost of shipping finished products to where they'll be sold *[1 mark]*.

Page 7 — Product Sustainability

Q1 Sustainability means not causing permanent damage to the environment *[1 mark]* and not using up finite resources *[1 mark]*.

Q2 E.g. the amount of greenhouse gases released into the atmosphere *[1 mark]* by making, using and eventually reusing, recycling or disposing of something at the end of its lifetime *[1 mark]*.

Q3 E.g. when a new product is designed so it can be easily taken apart at the end of its lifetime *[1 mark]*.

Page 9 — Product Sustainability and Social Issues

Q1 To work out the potential environmental impact of a product over its entire life cycle *[1 mark]*.

Q2 B *[1 mark]*

Q3 Any two from: e.g. providing them with a safe working environment / providing sufficient protective equipment (e.g. safety goggles) / providing training about how to work safely *[2 marks]*.

Page 11 — Products in Society

Q1 Any two from: e.g. it might allow the phone to be cheaper / better at its function / more efficient / nicer-looking *[2 marks]*.

Q2 So that you don't isolate or offend anybody in that culture *[1 mark]*.

Page 13 — Powering Systems

Q1 Advantage: e.g. nuclear power is reliable / nuclear fuel is quite cheap / nuclear power is clean/ produces very low levels of greenhouse gases *[1 mark]*. Disadvantage: e.g. nuclear fuel sources are finite / nuclear power plants cost a lot to build, maintain and decommission / the waste produced by nuclear power is very dangerous/difficult to dispose of / nuclear power always carries the risk of a major catastrophe *[1 mark]*.

Section Two — An Introduction to Materials and Systems

Page 15 — Properties of Materials

Q1 a) Materials that are fusible/have high fusibility have low melting points/easily turn into liquids *[1 mark]*.

b) Can be bent and shaped *[1 mark]*

c) Can be drawn into a wire *[1 mark]*

Q2 E.g. strong / tough / highly resistant to moisture *[1 mark]*.

Q3 E.g. metals are good thermal conductors / malleable *[1 mark]*.

Q4 Plastics are good electrical insulators *[1 mark]*.

Q5 Any two from: easily dyed / vulnerable to stains / dry slower *[2 marks]*

Q6 Any two from: e.g. strong / malleable / good conductor of heat / good conductor of electricity / low fusibility/ high melting and boiling point *[2 marks]*.

Q7 A mixture of two or more metals, or a metal mixed with one or more elements *[1 mark]*.

Page 17 — Paper, Board and Timber

Q1 The ink spreads out *[1 mark]*.

Q2 Paper has a weight of less than 200 gsm and board has a weight of more than 200 gsm *[1 mark]*.

Q3 High strength-to-weight ratio *[1 mark]*

This is important when moving heavy loads — you want the crate to be able to carry the load without breaking, but at the same time you don't want to have to carry a really heavy crate as well as what's in it.

Q4 a) E.g. oak / mahogany / beech / balsa / ash *[1 mark]*

b) E.g.
Oak: Used in interior panelling, flooring and furniture *[1 mark]*.
Mahogany: Used in furniture *[1 mark]*.
Beech: Used in toys and chairs *[1 mark]*.
Balsa: Used for modelling *[1 mark]*.
Ash: Used in tool handles *[1 mark]*.

Answers

Page 19 — Metals, Alloys and Polymers

Q1 A metal that doesn't contain iron *[1 mark]*, e.g. aluminium / brass / copper / tin / zinc *[1 mark]*.

Q2 Stainless steel doesn't rust/is corrosion-resistant *[1 mark]*. This is useful for outdoor products as they're often exposed to moisture which causes rust *[1 mark]*.

Q3 Any two from: thermosetting plastics are resistant to heat and fire but thermoforming plastics are not / thermosetting plastics are hard and rigid but thermoforming plastics are flexible / thermosetting plastics are non-recyclable/ cannot be remoulded whereas thermoforming plastics can be recycled/remoulded *[2 marks]*.

Q4 E.g. it is scratch-resistant *[1 mark]*.

Page 21 — Textiles

Q1 Any two from: natural fibres are biodegradable / recyclable / made from renewable materials *[2 marks]*.

Q2 E.g. elastane *[1 mark]* / polyester *[1 mark]*

Q3 a) E.g. cotton *[1 mark]*
b) Cotton is a cheap material *[1 mark]* and is also absorbent so it can be dyed *[1 mark]*.
c) E.g. cotton creases easily / shrinks when washed / dries slowly *[1 mark]*

Q4 filaments *[1 mark]*

Q5 By interlacing two sets of yarns *[1 mark]*.

Page 23 — Textiles and Manufactured Boards

Q1 The weft *[1 mark]* and the warp *[1 mark]*.

Q2 loom *[1 mark]*

Q3 E.g. warp-knitted fabrics keep their shape but weft-knitted fabrics can lose their shape / warp-knitted fabrics are less likely to 'ladder' than weft-knitted fabrics *[1 mark]*.

Q4 It's absorbent, so would become damaged by the moisture in the bathroom *[1 mark]*.

Page 25 — Electronic Systems

Q1 Input *[1 mark]*, process *[1 mark]* and output *[1 mark]*.

Q2 E.g. signals are what pass between blocks in a system *[1 mark]*. Blocks in a system change the signal in some way *[1 mark]*.

Q3 Boards with thin copper tracks *[1 mark]* that connect components in a circuit *[1 mark]*.

Q4 a) To change the resistance of the circuit (and therefore the current) *[1 mark]* based on some form of external factor (e.g. temperature) *[1 mark]*.
b) E.g. thermistor (temperature) *[1 mark]*, light-dependent resistor (light) *[1 mark]*, pressure sensor (pressure) *[1 mark]*.
c) E.g. *thermistor*: central heating system *[1 mark]*
LDR: automatic night light *[1 mark]*
Pressure sensor: car fuel system *[1 mark]*

Page 27 — Electronic Systems

Q1 E.g. a tiny, self-contained circuit that contains lots of components *[1 mark]*.

Q2 Any two from: e.g. microcontrollers can do the jobs of multiple integrated circuits, so systems containing microcontrollers can be simpler / microcontrollers can be reprogrammed for different functions whereas other integrated controllers cannot *[2 marks]*.

Q3 E.g. using a programming language (e.g. BASIC) to write a series of commands *[1 mark]*. Writing instructions in the form of a flowchart *[1 mark]*.

Q4 E.g. to set the timing of flashing lights/car indicators *[1 mark]* / to set the length of time that a microwave oven is on for *[1 mark]*.

Q5 The output will be on if both of the two inputs are on, otherwise the output will be off *[1 mark]*.

Q6 Any two from: e.g. lamp (light) / LED (light) / buzzer (sound) / speaker (sound) *[1 mark for naming two correct components, 1 mark if both output signals are correct]*

Page 29 — Mechanical Systems

Q1 a) E.g. moving backwards and forwards in a straight line *[1 mark]*. For example, a piston *[1 mark]*.
b) E.g. moving backwards and forwards in an arc *[1 mark]*. E.g. a swing *[1 mark]*.

Q2 E.g. a wheelbarrow *[1 mark]*

Q3 Gear ratio = 50 ÷ 10 = 5/1 (or 5:1 or 5) *[2 marks for correct answer, otherwise 1 mark for correct working]*

Page 31 — Mechanical Systems

Q1 a) a belt drive *[1 mark]*
b) It transfers movement from one rotating shaft to another *[1 mark]*.

Q2 a) Velocity ratio = 128 ÷ 32 = 4/1 (or 4:1 or 4) *[2 marks for correct answer, otherwise 1 mark for correct working]*
b) Output speed = 1200 ÷ 4 = 300 rpm *[2 marks for correct answer, otherwise 1 mark for correct working]*
c) e.g. a washing machine *[1 mark]*

Q3 A wheel with a grooved outer edge *[1 mark]*, and a cable/ rope/belt that sits in this groove *[1 mark]*.

Q4 a) E.g. a single pulley / a block and tackle *[1 mark]*
b) A single pulley changes the direction of the force required, so she can lift the load by pulling downwards which might be easier. / A block and tackle reduces the amount of force needed to lift the load *[1 mark]*.

Q5 a) For half a turn the follower won't move, then it will gently rise and fall *[1 mark]*.
b) To reduce friction *[1 mark]*.
c) pear cam *[1 mark]*

Answers

Page 33 — Developments in New Materials

Q1 Any two from: e.g. lightweight / strong / great conductor of heat / great conductor of electricity *[2 marks]*.

Q2 silver *[1 mark]*

Q3 Any two from: e.g. boats / some car bodies / kayaks / surfboards / PCBs *[2 marks]*.

Section Three — More about Materials

Page 35 — Selecting Materials

Q1 Any three from: e.g. functionality / availability / aesthetics / cost / environmental factors / social factors / ethical factors / cultural factors *[3 marks]*.

Q2 Any two from: it might be difficult to source / it may have to be sourced from further away / it is likely to be more expensive *[2 marks]*.

Q3 E.g. mahogany is an expensive material, so he'll need to sell it at a high price to make the money back that he spent on raw materials *[1 mark]*. It's a one-off product, so he cannot get any discounts on the price of the mahogany by buying in bulk *[1 mark]*.

Page 37 — Forces and Stresses

Q1 compression *[1 mark]*

Q2 tension *[1 mark]*

Q3 e.g. a guillotine / a die cutter *[1 mark]*

Q4 E.g. by trying to rotate the ends in opposite directions *[1 mark]*.

Page 39 — Scales of Production

Q1 One-off production, batch production, mass production *[1 mark for the correct order]*.

Q2 one-off production *[1 mark]*

Q3 E.g. you do one process on the whole batch, then you do another process on the whole batch *[1 mark]*.

Q4 Down time is the time between batches when you're not making anything *[1 mark]*. During this time, machines and tools may have to be set up differently/ changed for the next batch *[1 mark]*.

Q5 E.g. newspapers / magazines / cars *[1 mark]*.

Q6 a) mass production *[1 mark]*
b) It's easy to recruit people because they don't have to be highly skilled *[1 mark]*.

Q7 E.g. because it would be more expensive to keep stopping and restarting the process *[1 mark]*.

Page 41 — Quality Control

Q1 C *[1 mark]*

Q2 To check whether the size of a part is within the chosen tolerance *[1 mark]*.

Q3 a depth stop *[1 mark]*

Page 43 — Production Aids

Q1 a) A point where measurements are made from *[1 mark]*.
b) Any two from: e.g. they help to control accuracy / they're used to align CAM machines to the materials that they're operating on / they help to scale the size of something *[2 marks]*.
c) the datum/(0,0,0) *[1 mark]*

Q2 B *[1 mark]*

Q3 a) E.g. gets rid of the need for marking out / helps cut down on errors / makes sure that components are consistent / enables complex joints to be machined quickly and easily *[1 mark]*.
b) Because it would take him a long time to make a jig *[1 mark]*.

Page 45 — Production of Materials

Q1 a) crude oil *[1 mark]*
b) A process that separates crude oil out into chemicals called fractions *[1 mark]*.

Page 47 — More on the Production of Materials

Q1 a) C *[1 mark]*
b) Any two from: e.g. polyester / LYCRA® / nylon *[2 marks]*.

Q2 E.g. artificial fertilisers and pesticides are often used, which can pollute rivers and harm wildlife *[1 mark]*. Natural habitats may be cleared to be used as farm land *[1 mark]*.

Section Four — Paper and Board

Page 49 — Properties of Paper and Board

Q1 Recycled paper or board may contain toxic chemicals (which could enter the food) *[1 mark]*.

Q2 Any two from: e.g. printable / waterproof / airtight / strong *[2 marks]*.

Q3 To modify the properties of the paper/board *[1 mark]*.

Q4 As sheets or rolls *[1 mark]*.

Q5 A2 is 4 times larger than A4 (by area) *[1 mark]*.

Page 51 — Standard Components

Q1 Common fixings and parts that manufacturers buy instead of manufacturing them themselves *[1 mark]*.

Q2 drawing pins *[1 mark]*

Q3 E.g. comb binding / spiral binding / wiro binding *[1 mark]*.

Page 53 — Working with Paper and Board

Q1 a) To cut arcs and circles in card or paper *[1 mark]*.
b) To make tear-strips *[1 mark]*.
c) For precision cutting *[1 mark]*.

Q2 A laser cutter *[1 mark]*

Q3 To create an indent in a sheet of paper or board along which the material can be folded neatly and accurately *[1 mark]*.

Q4 The blades have to be made specially to match a net, so it can be expensive *[1 mark]*.

Page 55 — Printing Techniques

Q1 flexography *[1 mark]*

Q2 lithography *[1 mark]* and gravure *[1 mark]*

Q3 Digital printing is expensive per sheet but for short print runs it works out cheaper than setting up the plates for another printing process *[1 mark]*.

Q4 Cyan, magenta, yellow and black/key *[1 mark for two correct answers, 2 marks if all four are correct]*.

Page 57 — Paper and Board Finishes

Q1 a) E.g. business cards / menus posters *[1 mark]* because it will help them to last longer without getting damaged *[1 mark]*.

b) E.g. greetings cards / book titles / wrapping paper *[1 mark]* because it draws attention to a certain part of the product / gives the product a high-quality appearance *[1 mark]*.

Q2 a) It is where a shaped die is used to give a slightly raised impression on the surface of paper or board *[1 mark]*.

b) E.g. to add texture / draw attention to a particular area (e.g. a book title) / add a feeling of quality to a product *[1 mark]*.

Q3 E.g. a varnished surface can't be written on *[1 mark]*.

Q4 a) UV varnishing / UV coating *[1 mark]*.

b) It speeds up the printing process/ reduces production time *[1 mark]*.

Section Five — Wood, Metals and Polymers

Page 59 — Uses of Wood, Metals and Polymers

Q1 Any two from: e.g. it is strong / it is hard / it has an attractive appearance / it can be painted *[2 marks]*.

Q2 Any two from: e.g. it is strong / it doesn't easily rust / it has a high melting point / it is safe for use with food *[2 marks]*.

Q3 Any two from: e.g. it is strong / it has a uniform texture/it has no knots/grain / it can be shaped / it can have finishes applied to it *[2 marks]*.

Q4 its strength *[1 mark]*

Q5 a) annealing *[1 mark]*

b) First the metal is heated *[1 mark]*. Then it is left to cool slowly *[1 mark]*.

Page 61 — Stock Forms and Standard Components

Q1 E.g. steel / brass / stainless steel *[1 mark]*.

Q2 To join metal parts together *[1 mark]*.

Page 63 — More Standard Components

Q1 E.g. a flush/butt hinge *[1 mark]*

Q2 Advantage: e.g. they are very fast to use *[1 mark]*.
Disadvantage: e.g. the joints they produce are not as strong as glued joints *[1 mark]*.

Page 65 — Shaping Materials — Hand Tools

Q1 C *[1 mark]*

Q2 e.g. a bench plane *[1 mark]*

Q3 a) A — twist bit *[1 mark]*
B — flat bit *[1 mark]*

b) Drill bit A is used to used to drill small holes in wood, metal or plastic *[1 mark]*.
Drill bit B is used to drill large, flat-bottomed holes in wood and plastic *[1 mark]*.

c) a bradawl *[1 mark]*

Q4 Any two from: e.g. make sure the object you're working on can't move by holding it firmly in a bench hook, vice or clamp. / Keep your fingers out of the way. / Be aware of the exposed blade. / Always carry the saw safely. / Wear goggles *[2 marks]*.

Page 67 — Shaping Materials — Power and Machine Tools

Q1 a) e.g. planer *[1 mark]*
b) e.g. router *[1 mark]*

Page 69 — Shaping Techniques

Q1 To remove thin layers from a piece of material *[1 mark]*. To remove material to make a surface absolutely flat *[1 mark]*.

Q2 a) They are more malleable *[1 mark]*.

b) A sheet of metal is placed between a punch and a die *[1 mark]*. A hydraulic press moves the punch down and a large force is applied to the metal sheet *[1 mark]*. This pushes the sheet of metal into the die, which produces the required shape in the pressed metal sheet *[1 mark]*.

Q3 line bending *[1 mark]*

Page 71 — Moulding and Joining

Q1 E.g. a mould is put onto the vacuum bed, and a sheet of thermoforming plastic is clamped above it *[1 mark]*. The plastic is heated until it goes soft *[1 mark]*. The vacuum bed is lifted close to the heated plastic *[1 mark]*. The air is sucked out from under the plastic, forcing the plastic onto the mould *[1 mark]*. The moulded plastic is cooled and the vacuum bed is lowered *[1 mark]*.

Q2 tin and other metals *[1 mark]*

Page 73 — Treatments and Finishes

Q1 B *[1 mark]*

Section Six — Textiles

Page 75 — Fabrics and Their Properties

Q1 It's a good thermal insulator/ warm *[1 mark]* and is easy to wash *[1 mark]*.

Q2 E.g. cotton is abrasion-resistant *[1 mark]*, so it won't wear thin very easily/will be durable *[1 mark]*.

Q3 E.g. Work-wear for welders/ overalls / racing drivers' overalls / night clothes / children's pyjamas / fabric for soft furnishings *[1 mark]*.

Q4 A fabric made up of two or more layers that are stuck together *[1 mark]*.

Page 77 — Standard Components and Tools

Q1 2.7 m² *[2 marks for correct answer, 1 mark for correct working but incorrect answer]*.

Q2 a) Advantage: any one from: they're secure fastenings / quick and easy to attach / lay flat / don't add bulk / hard wearing / can be washed / quick to do up and undo *[1 mark]*.
Disadvantage: e.g. they can snag delicate fabrics *[1 mark]*.

Answers

b) Advantage: any one from: safe and soft / can be machine washed / hard-wearing *[1 mark]*. Disadvantage: any one from: hooks collect fibres over time and become less sticky / not very decorative *[1 mark]*.

Q3 Pinking shears *[1 mark]* because they cut the fabric with a zig-zag edge, which helps to prevent fraying *[1 mark]*.

Page 79 — Joining and Shaping Fabrics

Q1 To temporarily hold fabric in place while it is machine sewn *[1 mark]*.

Q2 E.g. a sewing machine uses two threads, one thread on the bobbin *[1 mark]* and one on the spool pin *[1 mark]*. The machine interlocks both threads together to form stitches *[1 mark]*.

Q3 To stop the edges from fraying *[1 mark]*.

Q4 a) E.g. sewing buttonholes / attaching pockets *[1 mark]*.

b) E.g. CAM embroidery machines / CAM knitting machines / CAM cutting machines *[1 mark]*.

Page 81 — Joining and Shaping Fabrics

Q1 a) French seam *[1 mark]* because there are no exposed raw edges to irritate the skin *[1 mark]*.

b) Flat-felled seam *[1 mark]* because it is a strong seam / stops edges from fraying *[1 mark]*.

Q2 E.g. piping *[1 mark]*

Q3 E.g. waistbands / cuffs *[1 mark]*

Q4 Advantage: any one from: it can create an interesting 3D effect / it adds warmth *[1 mark]*. Disadvantage: any one from: it requires a lot of material / it is time-consuming *[1 mark]*.

Page 83 — Dyeing

Q1 Any two from: the colours are brighter / it is easier and cheaper to make / exactly the same colour can be achieved again and again (which is more difficult with natural dyeing) *[2 marks]*.

Q2 A mordant is a chemical that fixes the colour of a dye to a fabric / makes it colour-fast *[1 mark]*. An example is salt *[1 mark]*.

Q3 E.g. batch dyeing *[1 mark]*

Q4 E.g. hot wax *[1 mark]*

Page 85 — Printing

Q1 They have a smooth surface for the dye to be applied to *[1 mark]*. They have no surface pattern to detract from the design *[1 mark]*.

Q2 E.g. Draw a pattern on a piece of wood then cut the background away, leaving the design raised / stick pieces of card or string onto a solid block *[1 mark]*.

Q3 To force the ink through the mesh of the screen and holes in the stencil *[1 mark]*.

Q4 Advantage: any one from: intricate patterns can be repeated quickly and accurately / it is much quicker than hand printing / large quantities of fabric can be printed at once / the screen-making process can be computer controlled, which makes it quicker *[1 mark]*. Disadvantage: any one from: setting up the machinery is expensive / making the screens can take a long time *[1 mark]*.

Section Seven — Electronic and Mechanical Systems

Page 87 — Properties of Components in Systems

Q1 Aluminium is more corrosion resistant than steel *[1 mark]* so will be more durable *[1 mark]*. Aluminium is less dense/lighter than steel *[1 mark]* so will improve vehicle acceleration, and fuel economy *[1 mark]*.

Q2 Any one from: e.g. strong/resists tensile forces/resists pulling forces / flexible *[1 mark]*.

Q3 anodisation *[1 mark]*

Page 89 — Standard Components in Systems

Q1 resistance = 390 ohms *[1 mark]* tolerance = 10% *[1 mark]*

Q2 Dual in Line package *[1 mark]*.

Q3 E.g. they can be reprogrammed many times so can be reused for different things *[1 mark]*.

Q4 Any two from: e.g. whether it's an extension or compression spring / width of the spring / number of coils / spring rate/how easy the spring is to stretch or compress *[2 marks]*.

Page 91 — Cutting, Drilling and Soldering

Q1 Any three from: plastic / wood / cardboard / fabric / some metals *[3 marks]*.

Q2 E.g. (it uses a lot of power to run, so) it can be expensive *[1 mark]*.

Q3 Components that have pins or wires poke through these holes which allows them to be soldered in place *[1 mark]*.

Q4 a) Solder provides an electrical connection between the component and the copper tracks *[1 mark]*. Soldering holds components in place on the circuit board *[1 mark]*.

b) E.g. a soldering iron *[1 mark]*

Page 93 — PCB Production and Surface Treatments

Q1 a) The layout of copper tracks is designed using CAD software and printed onto a transparent sheet of material (e.g. acetate) — this is the mask *[1 mark]*.

b) The mask is placed onto the blank PCB and the board is exposed to UV *[1 mark]*. The mask ensures the parts of the board that you want to remove are exposed to the UV light *[1 mark]*.

Q2 For such a small production run (10 PCBs), assembling components by hand is cheaper than using a machine, as he would have to buy (or hire) the machine first *[1 mark]*. He's using components that are pushed through holes in the board, which machines don't do very well *[1 mark]*.

Q3 Any two from: e.g. moisture / various chemicals / dust / large changes in temperature *[2 marks]*.

Q4 e.g. graphite / oil / silicone / grease *[1 mark]*

Answers

Q5 Any two from: e.g. vehicle engines / bike chains / locks / hinges *[2 marks]*.

Section Eight — Designing and Making

Page 95 — Looking at the Work of Designers

Q1 E.g. Marcel Breuer / William Morris / Philippe Starck / Louis Comfort Tiffany / Gerrit Rietveld / Ettore Sottsass *[1 mark]*.

Page 97 — Understanding User Needs

Q1 a) E.g. chest size / collar size / height *[3 marks]*.

b) E.g. the football shirt may be designed to fit 90% of the target market, not everybody *[1 mark]*.

Q2 E.g. it could have a large, easy-to-grip handle *[1 mark]*.

Page 99 — Design Briefs and Specifications

Q1 client *[1 mark]*

Q2 What kind of product is needed/wanted (and why) *[1 mark]*. How the product will be used *[1 mark]*. Who the product is for/the target market *[1 mark]*.

Q3 E.g. questionnaire / interview *[1 mark]*.

Q4 Any two from, e.g. how the product is made / how the product works / the size and weight of the product / how ergonomic the design is / how the product feels/looks/smells/tastes *[2 marks]*.

Q5 A list of conditions that the proposed product should meet *[1 mark]*.

Page 101 — Market Research

Q1 Any one of: e.g. it takes more time / it is harder to draw conclusions from the results *[1 mark]*.

Q2 E.g. to organise/sort/analyse data from market research / to present the data using charts and graphs *[1 mark]*.

Page 103 — Product Analysis

Q1 Any three from, e.g. function / form / ergonomics / competition and cost / sustainability / materials / manufacture *[3 marks]*

Q2 sub-assembly *[1 mark]*

Page 105 — Design Strategies

Q1 C *[1 mark]*

Q2 a) The prototype is tested (ideally by the target market) *[1 mark]*. The design is then evaluated, by examining the feedback/results of testing and identifying any problems with the design *[1 mark]*. Improvements are made to the design to fix these problems *[1 mark]*.

b) E.g. the same stages are repeated several times over in the same order *[1 mark]*.

Q3 To check that your design fits with the image they had in mind for the product / to gain useful feedback to improve future iterations of the design *[1 mark]*.

Q4 a) design fixation *[1 mark]*

b) Design strategies encourage you to look critically at your design and make improvements to it where necessary *[1 mark]*.

Page 107 — Exploring and Developing a Design Idea

Q1 a) Making practice versions of a design or parts of one *[1 mark]*.

b) It allows you to visualise your design in 3D *[1 mark]*, which allows you to spot and solve any potential problems with the design *[1 mark]*.

Q2 Toiles are early versions/models of a clothing design *[1 mark]*. They are used to work out the proportions and fit of a garment *[1 mark]*.

Q3 They don't require soldering *[1 mark]*, which means circuit components can easily be moved to change a circuit and create a new model/circuit design *[1 mark]*.

Q4 This is where you check the model against the design specification *[1 mark]*.

Q5 Developing a prototype *[1 mark]*.

Page 109 — Drawing Techniques

Q1 Drawing without using any equipment apart from a pencil or pen *[1 mark]*.

Q2 What something actually looks like in 3D — smaller in the distance and larger close up *[1 mark]*.

Q3 E.g. advantage: it's easier to get the dimensions right with isometric drawings *[1 mark]*. E.g. disadvantage: isometric drawings don't show things smaller the further away they are, so they look less realistic *[1 mark]*.

Q4 E.g. they aren't drawn to scale / they often use symbols / they often use simplified/unrealistic graphics *[1 mark]*.

Page 111 — More on Drawing Techniques

Q1 $15 \times 25 \times 5$ mm *[1 mark]*

Q2 Each separate part is moved out and drawn in line with the parts it's attached to *[1 mark]*. Dotted lines show where parts have exploded from/where they fit into the rest of the product *[1 mark]*.

Page 113 — Manufacturing Specification

Q1 a) A series of written statements, or working drawings and sequence diagrams, that tell the manufacturer exactly how to make a product *[1 mark]*.

b) Any three from: e.g. construction details / materials needed / equipment to be used / dimensions of each part / tolerances of parts / finishing details / quality control instructions / costings *[3 marks]*.

Q2 By putting the quality control checks in diamond-shaped decision boxes *[1 mark]*.

Q3 a) Because very few of the processes can happen at the same time *[1 mark]*.

b) 100 minutes *[1 mark]*

c) Allowing the paint to dry is the longest stage *[1 mark]*. It should take about 20 minutes to complete *[1 mark]*.

Answers

Page 115 — Developing Prototypes

Q1 Prototypes are full-size, fully-functioning products or systems *[1 mark]* made to test the product/system and its production methods *[1 mark]*.

Q2 E.g. to check that the manufacturing specification is correct *[1 mark]* and to help calculate manufacturing costs *[1 mark]*.

Q3 a) Any two from: e.g. clients / experts / potential customers *[2 marks]*.

b) E.g. feedback could provide additional suggestions to help modify the product *[1 mark]* to make it more marketable *[1 mark]*.

Q4 E.g. make a prototype with different fastenings and see if they look better *[1 mark]*.

Page 117 — Using Materials Efficiently

Q1 a) Making a mark on a material so you know where to cut, drill or assemble it *[1 mark]*.

b) E.g. a pencil / marking knife / marking gauge *[1 mark]*.

Q2 Repeats of the shape fit together without any gaps or overlapping pieces *[1 mark]*.

Q3 Finding an efficient arrangement of shapes to minimise waste during cutting and shaping *[1 mark]*.

Q4 Total area of stars
$= 150 \times 12 = 1800$ cm^2 *[1 mark]*
Area of board
$= 70 \times 50 = 3500$ cm^2 *[1 mark]*
Board wasted
$= 3500 - 1800 = 1700$ cm^2
[1 mark]

Page 119 — Working Safely

Q1 a) Any one from: e.g. protective gloves / apron / face shield *[1 mark]*.

b) E.g. chainmail gloves *[1 mark]*

Q2 So you don't breathe in toxic fumes/vapours *[1 mark]*.

Q3 a) Hazards *[1 mark]* and precautions/how to reduce the risk *[1 mark]*

b) To identify potential risks of the product to the end user *[1 mark]*, and make sure the product is designed to minimise these risks *[1 mark]*.

Q4 a) Standard components have already been tested for safety by the manufacturer, so you know they're likely to be safe to use in the product *[1 mark]*.

b) Prototypes can be used to carry out real-life simulations to test the product is safe to use *[1 mark]*.

c) CAD software can be used to simulate various types of tests (e.g. stress tests) to make sure the product is safe to use *[1 mark]*.

Index

TAR41